T0281482

Migration Imaging of the Transient Electromagnetic Method

Xiu Li · Guoqiang Xue · Changchun Yin

Migration Imaging of the Transient Electromagnetic Method

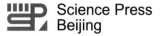
Science Press
Beijing

Springer

Xiu Li
Chang'an University
Xi'an
China

Guoqiang Xue
Institute of Geology and Geophysics
Chinese Academy of Sciences
Beijing
China

Changchun Yin
College of Geo-exploration Science
  and Technology
Jilin University
Changchun
China

ISBN 978-981-10-9688-4   ISBN 978-981-10-2708-6   (eBook)
DOI 10.1007/978-981-10-2708-6

Jointly published with Science Press, Beijing, 2016
ISBN: 978-7-03-050151-6 Science Press, Beijing

This Springer imprint is published by Springer Nature
The registered company is Springer Nature Singapore Pte Ltd.
The registered company address is: 152 Beach Road, #22-06/08 Gateway East, Singapore 189721, Singapore

Preface

Transient electromagnetic (TEM) method has been widely used in the exploration of mineral resources, environmental and engineering purposes, geological hazard survey, and other areas for its unique advantages. It has attracted much attention from geophysicists and geologists. With more and more mature or developing method from seismic being applied in TEM, the migration imaging of TEM data has become a hot research area in electromagnetic exploration. People are increasingly interested in methods such as pseudo-seismic inversion and interpretation, electromagnetic migration, or combination of migration and inversion. From the viewpoint of wave field, these methods not only expand and enrich research on TEM, but they also extract more meaningful information from the survey data that are more useful for imaging the underground targets but cannot be obtained via the conventional TEM method.

With enhancing the requirements of geophysicists in the resource exploration (deep mine and blind orebodies, etc.) and engineering survey (precise exploration), research and development of the multiple coverage technique for TEM method becomes necessary and practically significant. In addition, the multi-component array observation techniques can get abundant geoelectrical information and the technical research has also drawn attention. It is the direction and tendency for modern instruments to develop multi-component and multi-channel array transient EM systems that have strong anti-interference ability, high-power, timely data processing functions, versatile and intelligent. The development of high sensitivity, high performance and resolution, three-component sensors is also important part of the development of transient electromagnetic instruments.

On the basis of previous researches, the authors successively worked together with more than 20 graduate students to study pseudo-seismic interpretation of TEM data. We proposed the optimization algorithms for transformation from transient EM field to wave field and applied the regularization for the calculation of the inverse transformation, and successfully extracted virtual seismic wavelets. Experiments shown that using multi-aperture radiation field source we can obtain stronger magnetic field a single-aperture radiation field source. From observations of the primary field, we also find that multi-aperture field significantly improves the direction of the magnetic field. This confirms the existence of correlative stacking of multi-aperture transient EM field. We accomplished researches on pseudo-wavelet compression technique, applied Kirchhoff integral method to EM wave field migration, and achieved three-dimensional (3D) surface extension for TEM field. Starting from the characteristics of the EM field, we proposed the consecutive velocity analysis for TEM virtual wave based on the equivalent conducting plane and obtained initial results for TEM pseudo-seismic data interpretation.

This research monograph is intended to offer a reference to the readers who are interested in the technology of pseudo-seismic imaging and interpretation of TEM data. We divide the contents into seven chapters: Chapter 1 is the overview of development and prospects of TEM technology; Chapter 2 addresses the basic principle of wave field transform; Chapter 3 is to study the characteristics of pseudo-seismic field of TEM method; Chapter 4 introduces the synthetic-aperture and pseudo-wavelet compression technique; Chapter 5 discusses the 3D surface extension imaging technology of pseudo-seismic field; Chapter 6 is to study the velocity analysis of pseudo-seismic field; Chapter 7 demonstrates numerical simulation results on the migration imaging of theoretical synthetic models and examples of survey data from the authors' field work in the recent years.

The book is mainly composed by Dr. Xiu Li. Dr. Guoqiang Xue prepared Chaps. 1 and 4 and has proofread the book. Dr. Changchun Yin reviewed the book and made extensive revisions to the book. The graduate students Zhipeng Qi, Wenbo Guo, Junjie Wu, Tao Fan, Yin'ai Liu, Qiong Wu, Hongwei Zhu; graduate students in reading Huaifeng Sun, Jianbing Qian, and Yingying Zhang have also contributed to the research work and publication of this book. Nannan Zhou sorted out physical quantities of transient electromagnetic method and Weiying Chen checked the accuracy of the mathematical formulations, Hai Li checked the references, and Hai Li and Wuyang Li translated and proofread the directory. The authors acknowledge their contributions to this book.

This book is based on the author's projects supported by the National Natural Science Foundation of China and by enterprises in recent years. We must mention that the research and application of TEM in China starts relatively late, especially the theory and application of pseudo-seismic interpretation of TEM data is still at their early stages. Further researches are needed. We welcome readers to offer your kind comments and suggestions.

Xi'an, China   Xiu Li
November 2015

Contents

1 **Development and Prospect of Transient Electromagnetic
  Method** 1
  1.1 Overview on TEM Method 1
  1.2 The Development Direction of TEM 4
  1.3 Frontier Subject of TEM Pseudo-seismic Migration
      and Imaging 8
      1.3.1 TEM Imaging Based on Equivalent Time-Frequency
            Conversion 8
      1.3.2 TEM Imaging Technology Based on the Wave
            Field Conversion 10
      1.3.3 Reverse Time Migration and Imaging Method 11
  1.4 Research Progress of Pseudo-seismic Migration
      and Imaging in TEM 13
  References 14

2 **Theory and Technology of Full-Zone Wave Field
  Transformation** 17
  2.1 Theoretical Formula Construction 18
  2.2 Ill-Posedness of Inverse Wave Field Transformation Analysis 19
  2.3 Numerical Methods for Wave Field Transformation 20
  2.4 Preconditioned Conjugate Gradient Regularization for Inverse
      Wave Field Transformation 26
  2.5 Correlation Stacking Method for Extracting Pseudo
      Wave Field 31
  2.6 Effectiveness Test of the Algorithm 33
  References 37

3 **Property of TEM Pseudo Wave Field** 39
  3.1 Two-Layer Model Analysis 39
      3.1.1 Wave Field with Single Positive Peak 39
      3.1.2 Wave Field with Single Negative Peak 44

3.2   Three-Layer Model Analysis 49
      3.2.1   Q-Type Model with Double Positive Peaks 49
      3.2.2   H-Type Model with Positive–Negative Peaks 52
      3.2.3   K-Type Model with Negative–Positive Peaks 52
      3.2.4   A-Type Model with Double Negative Peaks 57
3.3   Wave Field Characteristics of Time-Domain EM Responses
      with Noise 57

**4   Synthetic Aperture Algorithms and Compression
      of Wavelet Width** 75
4.1   Imaging Method Based on Synthetic Aperture 75
4.2   Compression of Wavelet Width 77
      4.2.1   The Phenomenon of Waveform Dispersion 77
      4.2.2   The Reason of Waveform Dispersion 78
      4.2.3   Waveform Dispersion Compression Based
              on Deconvolution 79
      4.2.4   Model Calculations 82
Reference 85

**5   Surface Continuation and Imaging of TEM Based
      on Pseudo Wave Equations** 87
5.1   Establishment of Kirchhoff Diffraction Integral 87
5.2   Migration by Kirchhoff Integration (Surface Continuation) 93
5.3   Boundary Element Method for Wave Field Continuation 95
      5.3.1   Discretization of Kirchhoff Integration 95
      5.3.2   Analysis on Elements 95
      5.3.3   Total Matrix 97
      5.3.4   Integration Over Elements 98

6   Velocity Analysis of TEM Pseudo Wave Field 105
6.1   Velocity Modelling Based on Equivalent Conductive Plate
      Method 105
      6.1.1   Basic Theory of Equivalent Conductive Plate Method 106
      6.1.2   Approximate Calculation of TEM Field
              at the Surface of a Horizontally Layered Earth 108
      6.1.3   Optimized Extraction of Parameter $\bar{m}$ 110
6.2   Velocity Model for Single Observation Point 112
6.3   Continuous Velocity Analysis 114
      6.3.1   Weighted Interpolation Based on Global Distance 114
      6.3.2   Localized Linear Interpolation 115

7   Imaging of Theoretical Model and Field Examples 125
7.1   Model Calculations 125
      7.1.1   Layered Model 125
      7.1.2   Three-Dimensional Model 126

7.2   Examples with Field Data.   131
       7.2.1   Advanced Detection of Tunnel   131
       7.2.2   Detection of Goaf of Coal Mine   135

Chapter 1
Development and Prospect of Transient Electromagnetic Method

1.1 Overview on TEM Method

Transient Electromagnetic Method was first proposed by Soviet scientists in the 1930s that worked in far-field zone. The time-domain electromagnetic method with a dipole induced by a pulse current was proposed by Blau in 1933. Based on the similarity of electromagnetic wave at the formation interfaces of different conductivities with reflection signal of seismic wave, a lot of comparisons and experiments have been made. However, the frequency of transient EM signal induced by a pulse current was very low that made it difficult to achieve the resolution in identifying the reflection waves, resulting in TEM method be put aside. In 1950–1960s, Soviet scientists successfully completed one-dimensional (1D) forward modeling and inversion for TEM method and established the procedure for TEM field work and interpretations. TEM method began to enter the stage of practical use ever since. After the 1960s, people realized that time-domain electromagnetic sounding method can assume a much smaller transmitter–receiver offset than the expected exploration depth, TEM method underwent a rapid development. In Soviet Union, dozens of TEM teams were set up that made great achievements in oil and gas exploration. After that, the technologies like "short-offset," "late time," "near zone" were developed rapidly. In 1970–1980s, the short-offset method has been in the research and testing stage but not been widely used, while the long-offset TEM method has been used in the Western countries like the United States, particularly in geothermal exploration and investigation on crustal structure. During this time, people also conducted researches on 1D TEM forward modeling and inversion techniques. After 1980s, with the development of computer technology, European and American scholars published numerous papers on two-dimensional (2D) and three-dimensional (3D) forward modeling and inversion methods, using finite-difference, finite-element, integral equation algorithm, etc. The improvement on theory and understanding promotes the development and application of TEM method. The representative works of electromagnetic

© Science Press and Springer Nature Singapore Pte Ltd. 2017
X. Li et al., *Migration Imaging of the Transient Electromagnetic Method*,
DOI 10.1007/978-981-10-2708-6_1

theory were written by Kaufman and Keller (1983) in their monograph "Frequency-and Time-domain Electromagnetic Sounding" and by Nabighian (1992) in his edited book "Electromagnetic Method in Applied Geophysics" "Electromagnetic Geophysical Survey Method." Soviet scientist Бердчевский МН et al. (1985) proposed electromagnetic simulation of seismic wave migration methods. He put forward the generalized concept of the "migration imaging" and applied regularized migration and analytical migration in electromagnetic method. Subsequently, different kind of imaging techniques emerged. Among them, Macnae (1987), Nekut (1987), Eaton and Hohmamn (1987, 1989) have improved the conductivity imaging method, Christensen (2002) studied 1D imaging of TEM data, while Lee and Xie (1993) derived a 2D imaging interpretation method. Hoop (1996) suggested that there existed similarities between TEM method and reflection seismic exploration. Gershenson (1997) put forward a method using wave propagation characteristics to interpret time-domain electromagnetic sounding data. Kunetz (1972) studied the relationship between magnetotelluric and seismic elastic wave field. Lee and Memechan (1987) took advantage of the finite-difference method for 2D migration to complete electromagnetic data imaging based on the similarity between electromagnetic and seismic wave. Starting from the induced polarization theory, Berdichevsky and Zhadnov (1985) studied the effect of induced polarization on TEM method. They explained sign-changing phenomenon in EM responses of late time channels, and in the late 1980s they also investigated the numerical calculation of TEM responses for 3D polarizable bodies.

In China, the study on TEM method started in the 1970s. Niu (1986a, b, 1992), and Piao (1990), have worked earlier in this field. They applied airborne impulse EM system in the geological mapping and prospecting. Jiang et al. (1998) developed a transient electromagnetic pulse system and applied TEM in the exploration for good metal conductors. Niu et al. (1986, 1992) applied TEM method in mineral explorations and achieved good results. In cooperation with the Zhitong Institute for New Technologies, Changsha, China, they developed and produced the intelligent TEM instruments. Based on the Gaver-Stehfest inverse Laplace transform, Piao (1990) completed the forward calculation of TEM sounding of a ground source. With the digital filtering technology, Fang et al. (1993) completed the forward calculation of TEM responses for a large loop transmitter. The TEM sounding method with a large loop transmitter has been widely used in geothermal and groundwater explorations, engineering, and geological hazard surveys. They also applied TEM in the static correction of magnetotelluric soundings and achieved good results. Niu (1986a, b), Piao (1990), Fang et al. (1993), Jiang (1998), Yan et al. (2002), Bai (2003) conducted a lot of researches on TEM numerical simulation for 1D and 2D models, while Yin and Liu (1994), Chen (1998), Chen (1999), and Wang (2003) conducted researches on TEM forward modeling and inversion for 2D and 3D models. Based on the analysis of the similarity between electromagnetic plane waves and seismic waves, Wang (1985, 1986, 1990) systematically studied the pseudo-seismic migration imaging for TEM method. Based on the study of electromagnetic field, Yu and Wang (2001), Yu et al. (2003) put forward a modified finite-difference method for MT migration and imaging.

For transient electromagnetic sounding, Li et al. (2005a, b, c) carried out researches on imaging of differential conductance and proposed the optimization algorithm on TEM wave field. After transforming the EM field into wave field, they studied 3D imaging of the surface extension on the virtual wave of transient electromagnetic field. Chen (1999) carried out researches on the wave field transformation and he successfully obtained 2D geological images with a single interface for the synthetic data; Wang and Luo (2003) addressed the background of waveform broadening in the transformation of TEM wave field. Lv (1998) studied the reverse time migration (RTM) in 2D transient electromagnetic. Xue (2005) carried out researches on the ground TEM imaging and numerical calculation method; Yan et al. (2002) used the finite-difference method to calculate TEM responses and studied the joint inter-pretation of TEM sounding data in time-frequency domain. By analyzing the propagation characteristics of electromagnetic and seismic wave in inhomogeneous or layered media, Guo (2001) found that the diffusion of TEM field induced by a central loop source can be approximated by a spherical wave. In the late time, the spherical waves verge on plane waves, with similar propagation characteristics as seismic waves. Through a time-frequency transformation, the sequences for transmitting functions were calculated to perform pseudo-seismic migration and imaging for transient electromagnetic method.

There are many literatures on EM theory and applications published in China, mainly including "Theory on Electromagnetic Sounding" (Piao 1990), "Pulse Transient Electromagnetic Method and Its Application" (Niu 1986a, b) and "Theory on Time-Domain Electromagnetic Method" (Niu 1992), "Theory on Transient Electromagnetic Sounding" (Fang et al. 1993), "Practical Near-zone Transient Electromagnetic Method with a Magnetic Transmitting Source" (Jiang 1998), and "Theory and Application of Transient Electromagnetic Sounding" (Li 2002).

From the aspect of the instrument design and development, the Chinese research institutes have successfully developed many distinguished TEM instruments starting from the late 1980s. Geophysical and Geochemical Institute, Chinese Academy of Geological Sciences, developed WDC-1, WDC-2, and IGGETEM-20 TEM systems; X. Liu et al. from Geophysical and Geochemical Institute in Xi'an developed LC-1, EMRS-3 instruments, the transient electromagnetic systems with small loop but large transmitting current; Niu from Central South University together with Zhitong Institute for New Technologies developed SD-1, SD-2 TEM instruments. Changsha Baiyun Instruments Ltd. produced MSD-1 light-weighted TEM instruments. Wang (1999) from Beijing Institute of Geology and Mineral Resources, China Nonferrous Metals Group, developed TEMS-3S instrument. Lin et al. (2004) from Jilin University developed ATEM-2 instrument. Many explo-ration companies in the same period purchased advanced Western equipment, like the multifunctional electrical systems GDP-12, GDP-16, GDP-32 from the ZONGE in the US; V5, V6, V8 systems from Phoenix in Canada; PROTEM transient electromagnetic system EM37, EM47, EM57, EM67, etc. from Geonics in Canada; SIROTEM—II, III, Terra TEM instruments developed in Australia. Thus, the transient electromagnetic method has been promoted with a lot of successful applications and valuable research results.

With instruments becoming more and more intelligent and digital, TEM has been rapidly applied in engineering, environment, geological disaster survey since 1992, such as detection of coal gob area, collapse column; identification of underground faults; exploration of groundwater, mineral, petroleum, and coal mines; explorations for engineering purpose; hazard prediction in tunnels, and so on. Currently, the transient electromagnetic method has involved in almost all areas of exploration geophysics and achieved very good results.

1.2 The Development Direction of TEM

As an important electromagnetic exploration method, TEM method with its unique advantages (economic, nondestructive, rapid, rich in information, high resolution, etc.) is widely used in resource and engineering exploration. However, for the cases where a detailed geophysical investigation is needed, such as the advanced detection of tunnel for the highway and railway construction, detection of tomb structures in large scale, detailed hazard detection of dams, etc., resolution of commonly used TEM method is very limited, affecting the detections. To improve signal-to-noise ratio, one needs to increase the transmitter loop area or increase the dipole moment of EM transmitter. However, when transmitter loop area increases, the impact of the volume effect strengthens, which influences the resolution. Currently, theoretical research and instrument development are still at an early stage. Although theoretical research has solved 1D forward modeling and inverse problem, the studies on 2D and 3D have not yet reached the extent of applications, the instruments are mostly developed by improving the foreign instruments. Their applications are not widespread.

With the increasing requirements on resolution and precision, it becomes very urgent to conduct researches on the promising TEM method from two aspects. First, from the aspect of detection technology a survey mode based on multi-aperture array needs to be studied, with the aim to increase the resolution to the target body, while maintaining exploration depth but not increasing the volume effect. Second, profound researches on forward modeling and inversion, including development of new theories, are required to make TEM forward modeling and inversions systematic, to achieve the goal for TEM toward the 3D exploration and interpretation, and to solve the more complicated problems that require detailed investigations.

Since the 1970s, after 30 years of development, the researches on the technology and applications of TEM method have made great progress in China. A lot of monographs and articles on TEM theory and applications were published. Theory on data interpretation, techniques, and applications have been reported. However, there still exists large space for developments of TEM technology and applications.

Some experts believe that the TEM method develops slowly partly because the EM theory is very complicated, one needs broad knowledge on the equipment and the basic theory when using this technology. Since the production scale of domestic equipment is relatively small and the persistence are poor, coupled with limited promotion on applications, more and more domestic geophysical exploration units and research institutes purchase the multifunctional foreign equipment, restraining the domestic research on the method, the equipment and the technology. Compared with the developed countries, the application scale of TEM method in China and in the Western countries are roughly the same, while in other aspects, for example theoretical research and instruments development, are equivalent to the 1990s' level in Western countries. Forward modeling and inversion for complicated geoelectrical structures have just started. Jishan He (academician of China Academy of Sciences) noted that the 2D and 3D forward modeling, inversion, and imaging technology for electromagnetic method is an important direction for future theoretical researches. The TEM theory and applications currently still face the following problems:

(1) High-resolution data acquisition

As we know, transient electromagnetic method observes the time-varying secondary field induced by the underground media during the rest time of the primary field. The secondary field has the following characters: first, the signal dynamic range is large. The magnitude and attenuation speed of the signal depend on the time constant τ of underground media. For a conductor, the attenuation is slow, ranging from tens of microseconds to hundreds of milliseconds, the signal strength is generally from tens or hundreds of millivolts to a few tenths of microvolts. Second, the induced transient signal is broadband, with frequency ranging usually from n to $n \times 10^4$ Hz. Third, the influence from the interference noise is very serious, TEM generally works in the late time, and the survey signal is usually in the scale of microvolts. In this case, the natural and human interference noises are strong. The survey signal is actually the superposition of various kinds of signal. To improve the signal–to-noise ratio, it is necessary to improve the sensitivity of instruments or increase the power transmitter. Thus, how to sample and separate useful signal rich in information on the underground geoelectrical structures becomes the key to succeed in TEM survey.

In the current TEM technology, most instruments take the signal from the decay curve of the secondary induction field in a time window with logarithmically equal time interval, generally with 20–40 time channels, the information is very limited. GDP-32 multifunctional electrical survey system can take sample in arithmetically equal time interval. This greatly increases the sampling density. It can sample several hundreds to thousands of channel data and offer more geoelectrical information than the conventional sampling method, however, the interference noise comes in the late time channel that is stronger than useful information. Thus, it is important for exploration geophysicists to acquire and isolate the useful signal desired for resolving the underground geoelectrical structures.

For transient electromagnetic detection, regardless of long-offset or near-field observation, the current observations are mostly focusing on vertical component of magnetic field. Some scholars have suggested that the horizontal components should be observed simultaneously, but few successful applications have been reported. Thus, it is very meaningful and practical to study and develop array observation technology with multicomponents for the promotion of TEM applications.

(2) 2D, 3D Inversion theory

There are a lot of publications on TEM inversion theory and data interpretation worldwide, including the interpretation method based on a floating plate (interpretation method based on the apparent vertical conductance), the method based on the smoking ring theory, the method of pseudo-seismic migration and imaging, the method based on the time-frequency conversion, 1D and 2D forward modeling and inversion, and human–computer joint interpretations.

Joint interpretation of multicomponent, vector synthesis, and multiparameters has not been reported. The forward modeling and inversion of complex 2D and 3D models is not mature yet that is still in the stage of development. Therefore, the research on rapid interpretation of multicomponent, vector synthesis interpretation, the study on array observations and data processing method and on practical 2D and 3D forward modeling and inversion theory will greatly improve the resolvability and accuracy of interpretation, enrich TEM inversion theory and interpretation technology.

(3) Practicability

TEM method is widely used, with more and more attention being paid to it. Although this method bears unique advantages, it can still have difficulties when working in areas with specific geological and geomorphological conditions, such as in Qinling Mountains of China with harsh mountainous terrain and lush vegetation, TEM survey for mineral deposits is difficult, the layout of field work is very hard, the signal sampling is not easy, the rugged topography and lush vegetation lead even to possibility that the TEM method cannot be applied. Although many scholars engaged in experimental studies, marine transient electromagnetics still follow the traditional configuration and observation technology. This reduces the working efficiency and increases the difficulty to work. In addition, with increasing difficulty in detecting mineral deposits, the exploration in mining areas or surroundings with strong noise interference is needed. For engineering purpose, the detection of fine structures and complex geological sections are often required. There is an urgent need to develop methods and technologies with good anti-interference, great exploration depth, strong resolution, high-accuracy. Therefore, the development of cost-effective detection configurations and techniques, the high-accuracy interpretation theory and methods are the research direction for TEM.

The common depth point (CDP) multi-covering technology in seismic survey provides us new ideas on TEM research. A multicomponent TEM array observation will quickly acquire rich geoelectrical information. Mimicking CDP superposition in seismic exploration, a fast and practical interpretation method for array multi-component TEM data will greatly enhance the accuracy of transient electromagnetic interpretation, enrich TEM interpretation theory, and thus bring broad prospects for the development of transient electromagnetic method and applications (including airborne and marine TEM).

Therefore, we can summarize the development direction of TEM method in the future as follows:

(1) Theoretical research

In the study of 2D and 3D forward modeling and inversion for complex underground structures, attentions should also be paid to practical applications. Pseudo-seismic migration and imaging technology for TEM data and all kinds of signal separation technology from interference are noteworthy research directions. Both the Chinese and foreign scholars are searching a mathematical approach to convert transient electromagnetic field into wave field, which extracts the characteristics associated with the propagation of EM waves and suppresses or removes the characteristics related to dispersion and attenuation. If realized, solving the problem of transient electromagnetic field is converted into problems of solving the familiar wave equations. Seismic migration and imaging techniques, inversion via Born approximation, pulse spectrum inversion, tomographic inversions can be used in interpretation of TEM data. The interpretation techniques established on the base of motional similarity between EM wave propagating in a conductive medium and seismic wave propagating in elastic media has made significant progress in past years and achieved fruitful research results. This, to some extent, reduces nonuniqueness of EM solutions, expands the applications of electromagnetic methods. We can expect to extract more accurate information on underground earth from the survey data than the conventional inversion methods.

(2) Technologies

The research of TEM multiple coverage technique of pseudo-seismic is of big importance. In addition, the techniques with multicomponent and array survey should not be overlooked.

Due to the fact that a loop source is widely used in TEM method, study on techniques of high-resolution full-wave (both of on-time and off-time) detection, or improvement of traditional techniques will greatly increase resolution of TEM method both in vertical and in horizontal directions. This will enhance the accuracy of geological explanation and provide the technical support for a detailed survey for engineering purpose.

(3) Instruments

The multichannel, high-power, multifunctional, intelligent survey systems and high-sensitivity, high-performance sensors are difficulties in the development of EM instruments. The impact of market and Western equipment make it necessary to develop the domestic equipment in China with high- performance and resolution.

1.3   Frontier Subject of TEM Pseudo-seismic Migration and Imaging

1.3.1   TEM Imaging Based on Equivalent Time-Frequency Conversion

In recent years, theoretical and applied research of TEM prospecting for oil, geothermal, and mineral resources make continuous progress. For the detection of buried conductive bodies, TEM proves to be an effective technique. However, due to the complexity with EM field in stratified and dissipative media, the interpretation of survey data is currently still at very low level. Recently, people are working on the complex 2D and 3D models to describe more accurately the transient EM phenomenon in the earth and to get the high-resolution information on the shape and dimension of underground structures from the electromagnetic measurements.

TEM with a loop source is an electromagnetic detecting method of near-field observation. Due to the special characteristic of the transmitting source, the EM field propagating in the medium is a diffusion field. One cannot avoid this problem when studying the TEM imaging for a loop transmitter. The data conversion from a diffusive field to a plane wave field became a key issue.

Starting from large amount of forward modeling, comparison of model results for the two kinds of fields (diffusive field and plane field wave), and analysis of the propagation characteristics of two fields in underground media, Guo and Xue (2005) established the time-frequency equivalence between TEM sounding data to a plane wave field.

Based on error analysis of the synthetic modeling, accompanied by the analysis of field characteristics, we compare the penetration depth of two kinds of field and the consistency of underground electrical structures. After a detailed mathematical derivation, we come to the conclusion that the apparent resistivity from TEM sounding data can be converted into apparent resistivity for a plane wave field. There exists a conversion between the observation time and the frequency:

$$210/f = t, \tag{1.1}$$

where f is the frequency, t is the observation time. This relationship can be described in terms of time, speed, and depth of two fields propagating in the underground media. Diffusion velocity of EM field in conductive media is connected with the resistivity of underground media and time. At a time t after shutdown of a step wave, the relationship between the diffusion depth and the speed is expressed as

$$D(t) = \int_0^t V_d(\rho, t')\mathrm{d}t', \tag{1.2}$$

where $D(t)$ is diffusion depth, $V_d(\rho, t')$ is the diffusion velocity, ρ is resistivity of medium, t is time after the pulse is shut down.

In order to derive the propagation speed of TEM field and depth in underground media, we analyze the diffusion propagation mechanisms of TEM field. With the excitation of transmitting source, the eddy current is generated in the underground. After the step wave is turned off, the eddy current will not disappear immediately, but undergoes a transition process. The secondary induction field generated in the underground media goes through a process from small to big, to the peak value, and then decay to vanish. At certain depth in the underground, the initial magnetic field value is zero, with time it increases until it reaches a maximum value, and then decays to zero. EM field for certain frequency or at certain time distributes in any underground depth. From the perspective of geophysical prospecting, EM exploration depth is connected with the instrument sensitivity, geoelectrical conditions, noise level, and so on. Under ideal conditions, it is possible to detect the underground geological targets at a depth of several skin depths (or diffusion depth). However, in complex geological circumstances, it may not be possible to detect the targets buried in a skin depth (or diffusion depth). Overall, the apparent resistivity of a plane wave field and a diffusion field should have the same reaction to the underground electrical structures at the same depth.

In the 1D case, we consider that the skin depth and the diffusion depth are equal to EM detection depth. Thus, by making the two depths equal, we can obtain a formulation between the frequency and time [refer to Eq. (1.1)].

Xue et al. (2007) proposed the theory and numerical method for TEM imaging. They completed the transformation from a diffusion field to a plane wave field, from time-domain to frequency-domain. Based on this, one can calculate the impedance of the plane wave field from the converted apparent resistivity. Further from the impedance, one can establish equations system for obtaining the reflection coefficient sequence at each electrical layer. Finally, using a linear programming method, one can find the reflection coefficient sequence and draw the geoelectrical sections according to the reflection coefficients.

The numerical calculation of TEM imaging is divided into several steps: ①Various geoelectrical models are assumed to make the forward calculations. The apparent resistivities are calculated. Depending on the surface resistivities, the data at early time channels are corrected; ② Based on conversion relationship, the time delay is converted to frequency. The data for theoretical models or the apparent resistivities are transformed. The apparent resistivities of time-domain will be changed from the diffusion field into the plane wave field; ③ From the apparent resistivities of plane wave field, we obtain the impedance in the frequency-domain; ④ A equations system is built based on the impedance, and the sequence of reflection coefficients is solved by a linear programming algorithm; ⑤ An imaging profiles is created from the reflection coefficients.

1.3.2 TEM Imaging Technology Based on the Wave Field Conversion

Since the differential equation for TEM field is a diffusion one, it cannot be solved by the method for solving a wave equation. The so-called TEM wave field transformation means through a mathematical transformation, the time-domain TEM field that satisfies the diffusion equation, will be converted into the wave field that satisfies the wave equation. Then, using some more sophisticated seismic imaging techniques, we can obtain the properties and geometrical parameters of the exploration target.

In the 1970s and 1980s, the researches by Weidelt (1972), Kunetz (1972), Levy (1988), Lee (1987, 1989, Lee and Xie (1993) revealed that there exists an interesting mathematical correspondence between EM diffusion equation and seismic wave equation in the layered earth. The focus of their researches was to convert the model results for wave field into the time-domain EM responses. However, what interests the scholars is the inverse wave field transformation that can convert a time-domain field to wave field. This will be very helpful in applying the migration and more sophisticated imaging technology.

The transformation from the wave field to the time-domain field is expressed as

$$H_m(t) = \frac{1}{2\sqrt{\pi t}} \int\limits_0^\infty \tau e^{-\tau^2/4t} U(\tau) d\tau. \tag{1.3}$$

In the above equation, $H_m(t)$ is the TEM field, $U(\tau)$ is the virtual wave z, τ is the virtual time. This process is known as a inverse problem, which means that the time-domain field is known, while the wave field is to be solved. As we know that the inverse problem is generally linked to an ill-posed problem. A variety of inverse problems appear not only in the geophysics, but also in the mathematics itself. The wave field transformation is a typical first kind of Fredholm Operator Equation with a serious "ill-posed" feature. The concept "well-posed" and "ill-posed" was introduced by Hadamard in the early twentieth century to describe whether there exist a reasonable match between mathematical physics problem and the boundary conditions.

Diffusion equations satisfied by the TEM field describe diffusive characteristics of the induced eddy current. Migration and Inversion methods based on the diffusion equations are generally poor in identifying electrical interfaces. This requires to find a mathematical way to convert the TEM field into "wave field." His means that we need to extract the propagation characteristics from EM responses, to suppress or eliminate those features resulted from the dispersion and attenuation of EM waves during propagation. If we can convert the solution of TEM problem to a wave equation problem, we will be able to use techniques like seismic migration and imaging, Born approximation inversion, tomographic inversion in TEM inversion, and data interpretation.

Li and Xue (2005) presented some new theoretical analysis and numerical simulations of that transient electromagnetic diffusion-field response is transformed into a pseudo-seismic wavelet in engineering geology exploration. To simplify the integral equation used in the transformation, the integral range is separated into seven windows, and each window is compiled into a group of integral coefficients. Then, the accuracy of the coefficients is tested, and the calculated coefficients are used to derive the pseudo-seismic wavelet by optimization method. Finally, several geo-electric models are designed, so that model responses are transformed into the pseudo-seismic wavelet. The transformed imaginary wave shows that some reflection and refraction phenomena appear when the wave meets the electric interface. This result supports the introduction of the seismic interpretation in data processing of transient electromagnetic method

1.3.3 Reverse Time Migration and Imaging Method

(1) Stratton-Chu integral migration and imaging

After more than 10 years' systematic research, Zhdanov et al. (1988), Zhdanov and Booker (1993), Zhdanov and Li (1997) developed the RTM for time-domain TEM. They proposed the concept of electromagnetic migration and worked on 2D or 3D inverse problems based on RTM of EM field. Uses Stratton-Chu integration, the migration of TEM field can be expressed as

$$-\frac{1}{4\pi} \int_0^{t'} \iint_\Gamma \left\{ \begin{array}{l} (n \cdot E)\nabla G^* + (n \times E) \times \nabla G^* + (n \times H)\sigma H \\ G^* \end{array} \right\} \cdot dsdt. \tag{1.4}$$
$$= H(r', t)$$

Among them,

$$G^* = \frac{(\mu_0 \sigma)^{1/2}}{2\pi^{1/2}(t' - t)^{3/2}} \exp\left| -\frac{\mu_0 \sigma |r' - r|}{4|t' - t|} \right| u(t' - t) \tag{1.5}$$

with

$$u(t' - t) = \begin{cases} 1 & t \langle t' \\ 0 & t \rangle t' \end{cases} \tag{1.6}$$

In above equations, G^* represents the Green's Function, E and H are the electoral and magnetic field in Γ, n is a normal unit vector on Γ. Equation (1.4) shows that if we continue downwards the magnetic field observed at the earth's surface, we can get the magnetic field in the underground. After continuation, the EM field changes at the interface, thus we can identify the underground geoelectrical structures. Zhdanov et al. studied models with a single and multiple abnormal bodies and obtained good imaging results.

(2) Kirchhoff integration migration

Using an optimization algorithm, we convert TEM field into the "wave field," and we then use the wave field theory (namely the wave field is continued from the earth surface to the deep earth to realize the seismic migration and imaging) create the migration method for TEM to and interpret the TEM data. This means that we continue the EM survey data downward and produce the underground resistivity image from the EM responses.

Li (2005) proposed a numerical method that used Kirchhoff Integration method to carry out EM field migration. The propagation of wave field in the underground can be described by the following wave equation, i.e.,

$$\nabla^2 u - \frac{1}{v^2}\frac{\partial^2 u}{\partial t^2} = F, \tag{1.7a}$$

with the Kirchhoff Integral solution of

$$u(x,y,z,t) = -\frac{1}{4\pi} \oiint \left\{ [u]\frac{\partial}{\partial n}\left(\frac{1}{r}\right) - \frac{1}{r}\left[\frac{\partial u}{\partial n}\right] - \frac{1}{vr}\frac{\partial r}{\partial n}\left[\frac{\partial u}{\partial t}\right] \right\}dQ + \frac{F}{r_0}. \tag{1.7b}$$

The corresponding upward traveling wave fields $u(x,y,z,t)$ at an arbitrary depth position r, can be computed from the Kirchhoff–Helmholtz backpropagation formula

Assuming a zero-offset upward traveling wave of $G(x,y,z_0,t)$ that is the value of the wave field $g(x,y,z,t)$ at the earth surface $z = z_0$, induced by the "sources" at the reflection interfaces, we obtain from Eq. (1.7)

$$g(x,y,z,t) = -\frac{1}{4\pi} \iint \left\{ \frac{\partial}{\partial n}\left(\frac{1}{r}\right) - \frac{1}{r}\frac{\partial}{\partial n} - \frac{1}{vr}\frac{\partial r}{\partial n}\frac{\partial}{\partial t} \right\}G\left(\xi,\eta,\xi_0,t+\frac{r}{v}\right)dQ + \frac{F}{r_0}, \tag{1.8}$$

where $G\left(\xi,\eta,\xi_0,t+\frac{r}{v}\right)$ assumes a variable of $t+\frac{r}{v}$, because we consider the reverse process of wave propagation. We actually obtain the wave field in the underground from $g(\xi,\eta,\zeta_0,t)$—the value of $g(x,y,z,t)$ at the ground surface, so that we can determine reflection interfaces. This is the so-called the downward continuation of wave field.

3D boundary element technique can be introduced into the calculation of Kirchhoff Integration; the biggest advantage is to achieve 3D surface extension.

This is very difficult in seismic, because there exists $\frac{\partial G}{\partial n}$ in Eq. (1.8), while one cannot measure the vertical gradient t at the earth surface. On the contrary, one can easily measure the vertical gradient in TEM via a gradient sensor. Based on thorough research on RTM imaging in time-domain TEM, we can continue the TEM virtual wave field from the surface to the underground and establish the TEM migration and imaging method.

1.4 Research Progress of Pseudo-seismic Migration and Imaging in TEM

Xue et al. (2007) made new theoretical analyses and numerical experiments on pseudo-seismic wavelet transformed from TEM diffusion field for engineering exploration. The transformed virtual wave shows that reflections and refractions appear when the wave meets the electrical interfaces. This further supports the introduction of the seismic interpretation into the data processing of TEM method.

The directly transformed pseudo-wave has a dispersive and broad waveform so that it cannot clearly indicate the geological–electrical boundary of the underground objects. Xue et al. (2011) analyzed the reason why the waveform dispersion and presented a technique for sharping the transformed pseudo-seismic wave using a deconvolution method. Results show that the pseudo-seismic waves after deconvolution processes have sharper waveforms and narrower distribution zones. This can effectively control the waveform broadening phenomenon and is able to enhance the ability to recognize underground target using the TEM pseudo-seismic waves to determine the boundaries of underground objects.

Guo and Xue (2011) developed a new data-processing method that used the superposition to realize the integration of multi-aperture data as well as Kirchhoff migration and imaging. After the pseudo-wavelet extraction from TEM data, the traditional approach of profile-based multi-aperture synthesis has been developed for each survey station. Furthermore, the traditional single point approach was applied for multiple point coverage. The technology of synthetic aperture improves TEM resolution, rendering it possible to extract information from TEM data that cannot be obtained by conventional methods.

Xue et al. (2011) developed a sub-regularization algorithm to extract a virtual wavelet of the TEM field. According to the conventional designation of TEM recordings, the entire integration period is divided into seven time intervals. To avoid low accuracy in the calculations, high-density wave field data has been calculated based on the former sub-division. Therefore, the virtual wavelets can be extracted successfully using an optimized algorithm to obtain high-density integral coefficients for all time windows, and a satisfactory condition number of the coefficient matrix while taking a different channel number in each time period.

Xue et al. (2012) investigated a pseudo-seismic approach based on the so-called inverse Q-transform as an alternative way of processing TEM data. This technique transforms the diffusive TEM response into that of propagating waves obeying the standard wave equation. These transformed data can be input into standard seismic migration schemes with the potential of giving higher resolution subsurface images. Such images contain geometrical and qualitative information about the medium but no quantitative results are obtained as in model-based inversion techniques. These reconstructed images can be used directly for geological interpretation or in further constraining possible inversions. The results indicate that the resolution of the TEM data is significantly improved when compared with standard apparent resistivity plots, demonstrating that higher resolution 3D transient electromagnetic imaging is feasible using this method.

References

Bai DH, Meju, MA et al (2003) Numerical calculation of all-time apparent resistivity for the 215 central loop transient electromagnetic method. J Geophys 46(5):697–704

Chen MS (1999) Study on the transient electromagnetic (TEM) sounding with electric dipole. I basic principle. Coal Geol Explor 27(1):55–59

Christensen N.B (2002) A generic 1-D imaging method for transient electromagnetic data. Geophysics 67, 438–447

Eaton PA, Hohmann GW (1987) An eveluation of electromagnetic methods in the presence of geologic noise. Geophysics 52(8):1106–1126

Eaton PA, Hohmann GW (1989) A rapid inversion technique for transient electromagnetic sounding. Phys Earth Planet Inter 53(1989):394–404

Fang WZ, Li YG, Li X (1993) Principle of transient electromagnetic sounding. Northwestern Polytechnical University Press, Xi'an

Gershenson, M (1997) Simple interpretation of time-domain electromagnetic sounding using similarities between wave and diffusion propagation. Geophysics 62, 763–774

Hoop AT (1996) Transient electromagnetic vs. seismic prospecting—a correspondence principle. Geophys Prospect 44(6):987–995

Jiang BY (1998) Practical near-zone magnetic source transient electromagnetic prospecting. Geology Press, Beijing

Kaufuman AA, Kller GV, (eds) (1983) Frequency and time domain electromagnetic prospecting. Geology Press, Beijing (translated by Wang JM)

Kunetz G (1972) Processing and interpretation of magnetotelluric sounding. Geophysics 37(6):1005–1021

Lee KH (1989) A new approach to modeling the electromagnetic response of conductive medium. Geophysics 54(9):1180–1192

Lee KH, Xie G (1993) A new approach to imaging with low-frequency electromagnetic fields. Geophysics 58:780–786

Lee S, Memechan GA (1987) Phase-field imaging: the electromagnetic equivalent of seismic migration. Geophysics 52(5):678–693

Li X (2002) Theory and application of transient electromagnetic sounding. Shanxi Science and Technology Press, Xi'an

Li X (2005) Three dimensional curved surface continuation image based on TEM pseudo wave-field. Doctoral Dissertation Submitted to Xi'an Jiao Tong University, Xi'an

Li X, Xue GQ, Song JP, Guo WB, Wu JJ (2005a) The optimization of transient electromagnetic field-wave field conversion. Chin J Geophys 48(5):1185–1190 (in Chinese)

Li X, Song JP, Ma Y et al (2005b) The abstract of TEM signal based on the wavelet analysis. Coal Geol Explor 33(2):72–75

Li X, Xue GQ, Song JP et al (2005c) An optimize method for transient electromagnetic field-wave field conversion. Chin J Geophys 48(5):1185–1190

Lin J, Yu SB, Ji YJ (2004) New developments on study of TEM sounding system. In: Proceedings of the 20th Chinese geophysical society annual meeting. Xi'an Map Press, Xi'an, p 563

Lv GY (1998) Two-dimensional counter-time shift in transient electromagnetic method. Geophys Geochem Explor 22(2):139–142

Macnae JC (1987) An atlas of primary fields due to fixed transmitter loop EM source. Crone Limited Technical Note

Nabighian MN, (ed) (1992) Electromagnetic methods in applied geophysics. Geology Press, Beijing (translated by Zhao JX et al)

Nekut A (1987) Direct inversion of time-domain electromagnetic data. Geophysics 52, 1431–1435.

Niu ZL (1986a) Pulse induced transient electromagnetic method and its application. Zhongnan Industrial University Press, Changsha

Niu ZL (1986b) Time-domain electromagnetic theory. Zhongnan Industrial University Press, Changsha

Piao HR (1990) Principle of electromagnetic sounding. Geology Press, Beijing

Wang JY, Oldenburg Douglas, Levy Shlomo (1985) Magnetotelluric pseudo-seismic interpretation. Geophys Prospect Petrol 20(1):66–79

Wang JY, Fang S (1986) On magnetotelluric average velocity. Geophys Prospect Petrol 25(1): 79–84

Wang JY (1990) Magnetotelluric pseudo-seismic interpretation. Petroleum Industry Press, Beijing

Wang QY (1999) Study and manufacturing of the TEMS-3S transient electromagnetic sounding system. Geolog Explor Non-ferrous Metals 65(3):169–175

Wang HJ (2003) Study on 1-D forward and inversion of transient electromagnetic. Doctoral Dissertation Submitted to Xi'an Jiao Tong University, Xi'an

Wang HJ, Luo YZ (2003) Algorithm of a 2.5 dimensional finite element method for transient electromagnetic with a central loop. Chin J Geophys 46(6):855–862

Weidelt P (1972) The inverse problem of geomagnetic induction. J Geophys 38:257–298

Xue GQ (2005) Study on imaging method of loop-source transient electromagnetic method. Doctoral Dissertation Submitted to Xi'an Jiao Tong University, Xi'an

Xue GQ, Yan YJ, Li X (2007) Pseudo-seismic wavelet transformation of transient electromagnetic response in engineering geology exploration. Geophys Res Lett 34

Xue GQ, Yan YJ, Li X (2011) Control of the waveform dispersion effect and applications in a TEM imaging technique for identifying underground objects. J Geophys Eng 8, P 195

Xue GQ, BAI CY, LI X (2012) Extracting virtual reflection wave from TEM data based on regularizing method. Pure appl Geophys 69(7):1269–1282

Yan S, Chen MS, Fu JM (2002) Direct time-domain numerical analysis of transient electromagnetic fields. Chin J Geophys 45(2):275–284

Yin CC, Liu B (1994) The research on the 3D TDEM modeling and IP effect. Chin J Geophys 37(Suppl. I):486–492

Yu P, Wang JL (2001) Multi-parameter migration imaging of magnetotelluric data using finite difference method. Chin J Geophys 44(4):552–562

Yu P, Wang JL, Wu JS (2003) Summary and advances of magnetotelluric data imaging techniques. Prog Geophys 18(1):53–58

Zhdanov MS, Matusevich VU, Frenkel MA (1988) Seismic and electromagnetic migration. Nauka (in Russian)

Zhdanov MS, Booker JR (1993) Underground imaging by electromagnetic migration. In: 63rd Ann Internat Mtg Expl Geophys, Expanded Abstracts, pp 355–357

Zhdanov MS, Li WD (1997) 2D finite-difference time domain electromagnetic migration. In: 67th
    SEG EM2, vol 1, pp 370–373
Бердчевский МН, Zhdanov MC (1985) Interpretation of metaboli earth electromagnetic field.
    Geology Press, Beijing

Chapter 2
Theory and Technology of Full-Zone Wave Field Transformation

At present, the TEM pseudo-seismic interpretation has two approaches. One is to transform TEM diffusion field into MT planewave-like data through the approximate formula $t = (194–100)/f$ (t-ms, f-Hz) (Maxwell 1996). This was first used to correct the MT static shift according to TEM magnetic field sounding data so that seismic-like data can be obtained via the MT pseudo-seismic method (Sternberg 1985).

Another way is to transform TEM diffusion field into seismic-like data by using wave field transformation method. Because there is some relationship between the diffusion and wave equations, the similarities between wave propagation and diffusion propagation should also be there. Lee et al. (1987) formulated a complete imaging algorithm that is the EM equivalent of seismic migration. Based on the similarity of the Laplace transform for the diffusion and nondiffusion fields, Gershenson (1993) presented a simple transformation from the diffusion field into wave field; and this was proposed as a means of interpreting TEM sounding data. The diffusion equation for a field H in time domain t can be transformed into a wave equation for a field U in a time-like variable (Lee et al. 1989).

As a result, based on the relationship between diffusion equation and wave equation, TEM sounding data can be converted into "pseudo-seismic wavelet," and there will exist the certain relationship between converted imaginary wave and real seismic reflection wavelets. One can equivalently regard diffusion information as the seismic reflection signal which comes from underground interface.

However, the transforming equation will give rise to difficulties in the numerical calculation, so efficient numerical solutions still need to be developed. Based on the above-mentioned wave field transformation method, this chapter shows how to realize the wave field transformation and provides a more detailed description of the pseudo-seismic wavelet inversion.

© Science Press and Springer Nature Singapore Pte Ltd. 2017
X. Li et al., *Migration Imaging of the Transient Electromagnetic Method*,
DOI 10.1007/978-981-10-2708-6_2

2.1 Theoretical Formula Construction

In a homogeneous isotropic conductive medium, from the Maxwell equation, the partial differential equation of the magnetic field can be expressed as

$$\nabla \times E_m(r,t) = -\frac{\partial}{\partial t} B_m(r,t), \tag{2.1a}$$

$$\nabla \times H_m(r,t) = \sigma E_m(r,t) + \frac{\partial}{\partial t} D_m(r,t), \tag{2.1b}$$

$$\nabla \cdot H_m(r,t) = 0, \tag{2.1c}$$

$$\nabla \cdot E_m(r,t) = 0, \tag{2.1d}$$

where, $E_m(r,t)$ is the electrical field intensity, $H_m(r,t)$ is the magnetic field intensity, σ is conductivity, and r is the distance from the source to the field point. $D_m(r,t)$ is the electrical displacement vector.

And

$$B_m(r,t) = \mu H_m(r,t), \tag{2.1e}$$

$$D_m(r,t) = \varepsilon(r) E_m(r,t), \tag{2.2f}$$

where μ is the permeability and $\varepsilon(r)$ is the dielectric constant.

The initial and boundary conditions can be written as

$$H_m(r,0) = 0, \ H_m|_\Gamma = H_m(r_0,t), \ t > 0 \tag{2.2}$$

where Γ is the boundary of volume V at $r = r_b$.

We introduce $U(r,\tau)$ that satisfies the following wave equations

$$\nabla \times \nabla \times U(r,\tau) + \mu\sigma(r)\frac{\partial^2}{\partial \tau^2} U(r,\tau) = 0 \tag{2.3}$$

$$U(r,0) = \frac{\partial}{\partial q} U(r,\tau)|_{\tau=0} = 0 \tag{2.4}$$

$$U|_\Gamma = U(r_0,\tau), \quad \tau > 0 \tag{2.5}$$

where τ is the variable of pseudotime in the wave domain.

If $H_m(r,t)$ and $U(r,\tau)$ simplified as $H_m(t)$ and $U(\tau)$, based on the achievement from Lee et al. (1989),the unique relationship between the time-domain diffusion field $H_m(t)$ and the imaginary τ domain wave field $U(\tau)$ is given as follows

$$H_m(t) = \frac{1}{2\sqrt{\pi t^3}} \int_0^\infty \tau e^{-\tau^2/4t} U(\tau) d\tau \tag{2.6}$$

This transformation equation between the diffusion field $H_m(t)$ and the pseudo wave field $U(\tau)$ involves only time t and parameter τ which is independent of r.

According to Eq. (2.6), the value of the pseudo wave field $U(\tau)$ can be calculated using the surveyed value of the diffusion field $H_m(t)$. However, the equation is ill-posed and the unknown $U(\tau)$ is hidden in the integral equation.

2.2   Ill-Posedness of Inverse Wave Field Transformation Analysis

In the previous section, formula (2.6) is a typical first-class Fredholm operator equation, which gives a direct transformation from the wave field to time-domain diffusion field. However, time-domain diffusion field is known and corresponding unknown wave field need to be computed which makes the problem "ill-posed."

Assuming that ρ_F and ρ_U are the dimensions of the space F and U, respectively, the operator $A : F \rightarrow U$ is a linear or nonlinear mapping of F to U. The inverse problem in formula (2.6) can be written in the form of first-class operators, i.e.,

$$Az = u \quad z \in F \quad u \in U. \tag{2.7}$$

In above equation, A can be an integral operator. Thus, we make the definition: the problem in Eq. (2.6) is well posed, if it simultaneously meets the following three conditions:

1. $\forall u \in V$, there exists $z \in F$ that satisfies Eq. (2.7) (existence of solution);
2. Assuming $u_1, u_2 \in U$, z_1 and z_2 are respectively the solutions of Eq. (2.7), if $u_1 \neq u_2$, then $z_1 \neq z_2$ (uniqueness of solution);
3. Solutions are stable in the space domain (F, U) (stability of solution). This means that if $\forall \varepsilon > 0$, then $\delta(\varepsilon) > 0$. As long as

$$\rho_{U(u_1,u_2)} < \delta(\varepsilon) \quad (u_1, u_2 \in U), \tag{2.8}$$

we have

$$\rho_{F(z_1,z_2)} < \varepsilon \quad (Az_1 = u_1, Az_2 = u_2). \tag{2.9}$$

Conversely, if at least one of the above three conditions cannot be satisfied, we call it ill-posed.

From above discussion, we know that the existence and uniqueness of solutions depend on algebraic features of the space $(F,\ U)$ and the operators A, namely whether the operator is a surjection or one-to-one mapping (injection); the stability depends on the topological properties of space, namely whether the inverse operator A^{-1} is continuous. Thus, a well-posed and ill-posed problem is connected not only with the operator and its definition, as well as value domain, but also with dimension of the space.

Moreover, the above three conditions have practical significance. First of all, for practical problems, we hope that the solution exists and is unique. More important is that the field data actually are really not accurate. The "accurate" data are generally unknown, a small error in the original data can lead to a serious deviation from the true solution, leading to meaningless results. It is obvious that when dealing with geophysical issues, the well-posedness of an operator A is very important. To judge whether an operator is well-posed or not, we can take the following theorem: Assuming that $A : F \to U$ is completely continuous operator (compact operator), F, U are all Banach spaces. If A^{-1} exists, then for the case anyone of F and U has an infinite dimension, A must be an unbounded operator and is not continuous on U.

From this theorem, we know that even if the first class of integral equations with a completely continuous operator has a solution, the solution is unstable. Furthermore, when the value domain is not closed (because the error in the observations often leads to data beyond the value domain of the operator), the equation has no classical solutions. Thus, the first class of integral equations (including Fredholm and Volterra equations) is generally ill-posed.

2.3 Numerical Methods for Wave Field Transformation

For a homogeneous whole space, Eq. (2.6) can be simplified so that a simple relationship between the diffusion and wave fields can be determined. Consequently, a suitable numerical scheme must be used to solve the problem. In this section, to obtain the solution of Eq. (2.6), we will seek a group of integral coefficients to linearize the integral functions.

According to the duality in EM theory, each component of TEM field satisfies Eq. (2.6) of wave field transformation. Without loss of generality, we take $f(x, y, z, t)$ for TEM field components, $u(x, y, z, \tau)$ for the component of the pseudo-seismic field $U(\tau)$. Then, the numerical integration of (2.6) can be written as

$$f(x, y, z, t) = \frac{1}{2\sqrt{\pi t^3}} \int_0^\infty \tau e^{-\frac{\tau^2}{4t}} u(x, y, z, \tau) \mathrm{d}\tau. \qquad (2.10)$$

If the transient electromagnetic field $f(x, y, z, t)$ is known, the corresponding pseudo wave field $u(x, y, z, \tau)$ can be obtained by inverse transformation of (2.10). Further, the numerical integration of (2.10) can be written as

$$f(x, y, z, t_i) = \sum_{j=1}^{n} u(x, y, z, \tau_j) a(t_i, \tau_j) h_j, \qquad (2.11)$$

where h_j is the integral coefficient,

$$a(t_i, \tau_j) = \tau_j e^{-\frac{\tau_j^2}{4t_i}} \qquad (2.12)$$

is the kernel function that increases with τ until it reaches extreme value and then rapidly decays to zero (Fig. 2.1). Twenty kernel functions function curves for different time are shown in Fig. 2.1a, b. It is shown clear that for different time, the kernel functions curve have basically the same shape. However, the magnitude of kernel functions and dynamic range of the virtual time τ change a lot which indicate the difficulty in the inverse transform of the wave field.

Basically, there are several discretization ways from Eqs. (2.10) to (2.11). We compare some conventional discretization methods, such as equidistant discretization, logarithm equidistant discretization, discretization of equal area, and discretization of equal height, the virtual time distribution for different discretization methods are shown in Fig. 2.2. Where (a) shows the discretization result of equal area; (b) shows logarithm equidistant discretization; (c) shows the discretization of equal height; (d) shows linear equidistant discretization.

Fig. 2.1 Kernel functions for different time. **a** $t = 1$–100 μs, **b** $t = 0.1$–10 s

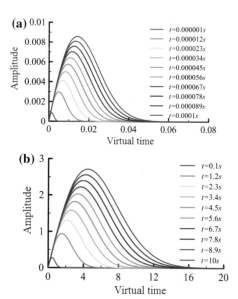

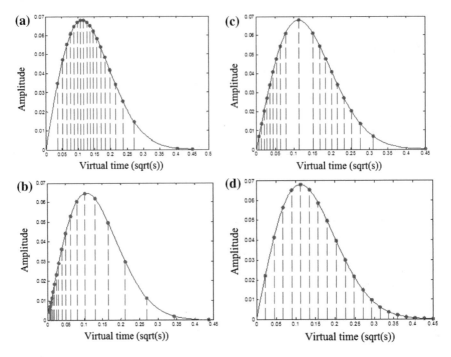

Fig. 2.2 Virtual time distribution for different discretization methods. **a** discretization of equal area; **b** logarithm equidistant discretization; **c** discretization of equal height; **d** linear equidistant discretization

According to different discretization methods, we select the virtual time and integrate numerically. The comparison of the accuracy of numerical integrations and matrix condition numbers are given in Table 2.1. We choose the equidistant discretization method for the smallest mean square error and best conditions number of the coefficient matrix.

The key for the calculation of Eq. (2.11) is to obtain a set of h_j, $\tau_j (j = 1, 2, \ldots n)$ to best satisfy Eq. (2.11). For this purpose, we

Table 2.1 Comparison of different discrete way

Discritization method	Mean square error (%)	Condition number of coefficient matrix cond(A)
Linear equidistant	2.53	7.58E+18
Logarithmic equidistant	4.92	1.05E+19
Discretization of equal area	6.94	3.25E+20
Discretization of equal height	8.74	1.53E+19

We also can rewrite Eq. (2.11) as follows:

$$H_m(t_i) = \sum_{j=1}^{n} U(\tau_j) a(t_i, \tau_j) h_j \tag{2.13}$$

where $a(t_i, \tau_j) = \dfrac{1}{2\sqrt{\pi t_i^3}} \tau_j e^{\frac{\tau_j^2}{4t_i}}$, h_j is the integral coefficient.

If the integral coefficients are known, we can obtain the "wave field" value $U(\tau_j)$ from the surveyed transient electromagnetic field value $H_m(t)$ by performing an inverse transformation of Eq. (2.13). Consequently, the integral coefficient h_j is key parameter to be solved.

If let $u(\tau) = 1$, Eq. (2.6) have the following form

$$H_m(t) = \frac{1}{2\sqrt{\pi t^3}} \int_0^{\infty} \tau e^{-\tau^2/4t} d\tau \tag{2.14}$$

According to the special integral

$$\int_0^{\infty} x e^{-ax^2} dx = \frac{1}{2a} \tag{2.15}$$

Equation (2.14) becomes

$$H_m(t) = \frac{1}{2\sqrt{\pi t^3}} \int_0^{\infty} \tau e^{-\tau^2/4t} d\tau = \frac{1}{\sqrt{\pi t}} \tag{2.16}$$

and the right-hand term of Eq. (2.13) can be written as

$$\sum_{j=1}^{n} a(t_i, \tau_j) h_j = \frac{1}{\sqrt{\pi t_i}} \tag{2.17}$$

Integral coefficients can be obtained from Eq. (2.17) using optimization method. We rewrite Eq. (2.17) as a matrix equation:

$$\begin{aligned} T_1 &= a_{11} h_1 + a_{12} h_2 + \cdots + a_{1N} h_N \\ T_2 &= a_{21} h_1 + a_{22} h_2 + \cdots + a_{2N} h_N \\ \cdots \quad &\cdots \quad \cdots \quad \cdots \quad \cdots \\ T_M &= a_{M1} h_1 + a_{M2} h_2 + \cdots + a_{MN} h_N \end{aligned} \tag{2.18}$$

where $T_i = \frac{1}{\sqrt{\pi t_i}}$, $a_{i,j} = a(t_i, \tau_j)$.

Because the temporal scope of transient electromagnetic field data is wide from several tens μs to several tens ms, the number of matrix coefficients will

Table 2.2 List of seven time range

Time range number	Time range
1	32.5–80 μs
2	80–325 μs
3	325–800 μs
4	900 μs–2.1 ms
5	2.4–8.7 ms
6	8.7–27 ms
7	27–81 ms

become excessive large. In order to overcome this problem, the entire time range (32.5 μs–81 ms) is separated into seven time ranges (shown in Table 2.2) so as to calculate the coefficients by solving Eq. (2.17) for each window.

For every time window, the n in Eq. (2.17) is selected as 20, thus 20 integral coefficients ($h_1 - h_{20}$) can be obtained by solving Eq. (2.17). All results of the acquired integral coefficients are listed in Tables 2.3, 2.4, 2.5, 2.6, 2.7, 2.8 and 2.9.

Table 2.3 The first time range coefficiency

i, j	Time/ti (s)	τ_j	h_j	i, j	Time/ti (s)	τ_j	h_j
1	0.0000325	0.00687374	0.0012802	11	0.0000575	0.0151322	0.00117166
2	0.000035	0.00713483	0.0012879	12	0.00006	0.0158077	0.00116311
3	0.0000375	0.00790998	0.00128198	13	0.000063	0.0170941	0.00119561
4	0.00004	0.00841461	0.00128879	14	0.000065	0.0187315	0.00126428
5	0.0000425	0.00907891	0.00129713	15	0.0000675	0.0222362	0.00160694
6	0.000045	0.00978989	0.00128997	16	0.00007	0.0281634	0.00305347
7	0.0000475	0.0107273	0.0012588	17	0.0000725	0.0282789	0.0030634
8	0.00005	0.0116837	0.0012082	18	0.000075	0.0282797	0.00306347
9	0.0000525	0.0130093	0.00119407	19	0.0000775	0.0282839	0.00306381
10	0.000055	0.0134242	0.00118834	20	0.00008	0.0282843	0.00306385

Table 2.4 The second time range coefficiency

i, j	Time/ti (s)	τ_j	h_j	i, j	Time/ti (s)	τ_j	h_j
1	0.00008	0.0103487	0.00474251	11	0.000209	0.0316496	0.00242758
2	0.000093	0.0111039	0.0042478	12	0.000222	0.0335201	0.00265559
3	0.000106	0.0132478	0.00367154	13	0.000235	0.035652	0.00287931
4	0.000119	0.0164748	0.00269769	14	0.000248	0.0386983	0.00315865
5	0.000132	0.0194231	0.00195261	15	0.000261	0.0413146	0.00341179
6	0.000144	0.0218342	0.00185143	16	0.000273	0.0563658	0.00553772
7	0.000157	0.023965	0.00180551	17	0.000286	0.0567938	0.00557217
8	0.00017	0.0259851	0.00185982	18	0.000299	0.056794	0.00557219
9	0.000183	0.02794	0.00192987	19	0.000312	0.056794	0.00557219
10	0.000196	0.0298215	0.00218228	20	0.000325	0.056794	0.00557219

Table 2.5 The third time range coefficiency

i, j	Time/ti (s)	τ_j	h_j	i, j	Time/ti (s)	τ_j	h_j
1	0.000325	0.0215624	0.00540507	11	0.000575	0.0638756	0.00406249
2	0.00035	0.023449	0.00553082	12	0.0006	0.0758292	0.00489693
3	0.000375	0.0258422	0.00554672	13	0.00063	0.0840326	0.00549048
4	0.0004	0.0286992	0.00553888	14	0.00065	0.0846608	0.00551579
5	0.000425	0.0321007	0.0054782	15	0.000675	0.0847099	0.00551762
6	0.00045	0.0360844	0.00513157	16	0.0007	0.0847178	0.00551792
7	0.000475	0.0406078	0.00462507	17	0.000725	0.0847199	0.00551799
8	0.0005	0.0455316	0.00427416	18	0.00075	0.0847206	0.00551802
9	0.000525	0.0507573	0.00404108	19	0.000775	0.0847209	0.00551803
10	0.00055	0.0565126	0.0040015	20	0.0008	0.0847211	0.00551804

Table 2.6 The forth time range coefficiency

i, j	Time/ti (s)	τ_j	h_j	i, j	Time/ti (s)	τ_j	h_j
1	0.0009	0.0342904	0.00808577	11	0.00154	0.0937619	0.00661376
2	0.000968	0.0373135	0.00808811	12	0.0016	0.10293	0.00689047
3	0.00103	0.0404446	0.00811089	13	0.00166	0.116617	0.0070268
4	0.00109	0.0444847	0.00812134	14	0.00173	0.132342	0.00889207
5	0.00116	0.0494548	0.00804132	15	0.00179	0.137734	0.0101515
6	0.00122	0.0552661	0.0079612	16	0.00185	0.140495	0.0104658
7	0.00128	0.0617234	0.00750109	17	0.00192	0.140515	0.0105672
8	0.00135	0.0686295	0.00724702	18	0.00198	0.140523	0.0106677
9	0.00141	0.0759422	0.00686272	19	0.00204	0.140523	0.0106677
10	0.00147	0.083962	0.00664923	20	0.0021	0.140523	0.0106677

Table 2.7 The fifth time range coefficiency

i, j	Time/ti (s)	τ_j	h_j	i, j	Time/ti (s)	τ_j	h_j
1	0.0024	0.0636036	0.0180072	11	0.00572	0.187814	0.0120336
2	0.00273	0.0702456	0.0181402	12	0.00605	0.211214	0.0138898
3	0.00306	0.0784513	0.0179162	13	0.00638	0.245295	0.0174021
4	0.00339	0.0882636	0.017208	14	0.00671	0.266237	0.0188688
5	0.00373	0.0996682	0.0159979	15	0.00704	0.269226	0.0189976
6	0.00406	0.112467	0.0144493	16	0.00737	0.269625	0.019013
7	0.00439	0.126229	0.0128946	17	0.00771	0.269709	0.0190162
8	0.00472	0.140474	0.0120819	18	0.00803	0.269734	0.0190172
9	0.00505	0.155019	0.0112268	19	0.00837	0.269744	0.0190175
10	0.00538	0.170303	0.0112284	20	0.0087	0.269748	0.0190177

Table 2.8 The sixth time range coefficiency

i, j	Time/ti (s)	τ_j	h_j	i, j	Time/ti (s)	τ_j	h_j
1	0.0087	0.120552	0.0441067	11	0.01833	0.32108	0.0156714
2	0.00966	0.136986	0.0398148	12	0.01929	0.34341	0.0179491
3	0.01062	0.15502	0.0359394	13	0.02026	0.36833	0.0217686
4	0.01159	0.174284	0.0309627	14	0.02122	0.399271	0.0274332
5	0.01255	0.194441	0.0256392	15	0.02218	0.442436	0.0358438
6	0.01352	0.215179	0.0207631	16	0.02315	0.490221	0.0435548
7	0.01448	0.23624	0.0169477	17	0.02411	0.509749	0.0460162
8	0.01544	0.257416	0.0155199	18	0.02507	0.515195	0.0460462
9	0.01641	0.278576	0.0145354	19	0.02603	0.515195	0.0460462
10	0.01737	0.299713	0.0146745	20	0.027	0.515195	0.0460462

Table 2.9 The seventh time range coefficiency

i, j	Time/ti (μs)	τ_j	h_j	i, j	Time/ti (μs)	τ_j	h_j
1	0.027	0.198459	0.0867648	11	0.05542	0.599458	0.0346452
2	0.02984	0.237216	0.0775172	12	0.05826	0.640064	0.0361105
3	0.03268	0.277437	0.0642214	13	0.0611	0.68172	0.0384676
4	0.03552	0.318166	0.0507675	14	0.06395	0.72524	0.0426428
5	0.03836	0.358856	0.0401601	15	0.06678	0.800949	0.0549936
6	0.04121	0.399315	0.0336501	16	0.06963	0.845541	0.0644859
7	0.04405	0.439544	0.0309344	17	0.07247	0.891042	0.0743667
8	0.04689	0.479588	0.0308258	18	0.07531	0.927646	0.0814274
9	0.04973	0.519498	0.0319624	19	0.07815	0.967922	0.0873604
10	0.05257	0.559383	0.0333534	20	0.081	0.967922	0.0873604

2.4 Preconditioned Conjugate Gradient Regularization for Inverse Wave Field Transformation

In the aforementioned wave field transformation, the optimization method is applied twice, it can minimize the number of integral coefficients and sampling points, while ensuring the accuracy of calculations. It can also reduce the order of equations and improve the ill-posedness of the equations system. Considering the discrete form (2.13) of wave field transformation where the condition number of the coefficient matrix is large, the single method of regularized conjugate gradient or the preconditioned conjugate gradient cannot well solve the problems. Therefore, we combine the two methods to the precondition regularized conjugate gradient method for solving this problem. Equation (2.13) can be casted into the matrix equation

$$AU = F, \tag{2.19}$$

where $A = [a_{ij} h_j]_{m \times n}$ and $U = [u_j]_{n \times 1}$ are the virtual wavelets, $F = [f_i]_{m \times 1}$ is the TEM field.

In order to use the conjugate gradient iteration method, Eq. (2.19) is reformulated as

$$A^T A U = A^T F. \tag{2.20}$$

In above equations, as long as A is column full rank matrix, $A^T A$ will be symmetric and positive-definite, so we can use the conjugate gradient method. The condition number of $A^T A$ is bigger than that of A, creating more serious ill-posed problem in Eq. (2.20). To further reduce the condition number of matrix and to improve the ill-posedness, we need to enforce the precondition of the coefficient matrix before the regularized conjugate gradient procedure. An over-relaxation preconditioned method is used to construct the preconditioning matrix, because the symmetric over-relaxation precondition is a more effective preconditioning method. The preconditions factor is easy to obtain, and the condition number of matrix can be reduced.

Assumed a matrix

$$S = L^T L + \alpha D^T D \tag{2.21}$$

can be decomposed into

$$S = M - N, \tag{2.22}$$

where

$$M = \frac{1}{\omega (2 - \omega)} \left[(K + \omega C_l)^{-1} K^{-1} (K + \omega C_u) \right], \tag{2.23}$$

$$N = \frac{1}{\omega (2 - \omega)} \left[((1 - \omega) K + \omega C_l)^{-1} K^{-1} ((1 - \omega) K + \omega C_u) \right], \tag{2.24}$$

While K, C_l, and C_u are, respectively, the diagonal elements of S, Lower triangular elements and upper triangular elements, ω is a parameter in the range of $(0, 2)$. Thus, we can select the precondition matrix P as

$$P = (K + \omega C_l)^{-1} K^{-1} (K + \omega C_u). \tag{2.25}$$

Mathematically, it has proved that after the over-relaxation preconditioning, the condition number of matrix falls to the square root of the original one.

Assuming that a preconditioning factor $M(v)$ for matrix $A(v)$ is constructed in Eq. (2.23), the matrix $M(v)^{-1} A(v)$ is close to the unit matrix, so that the iteration

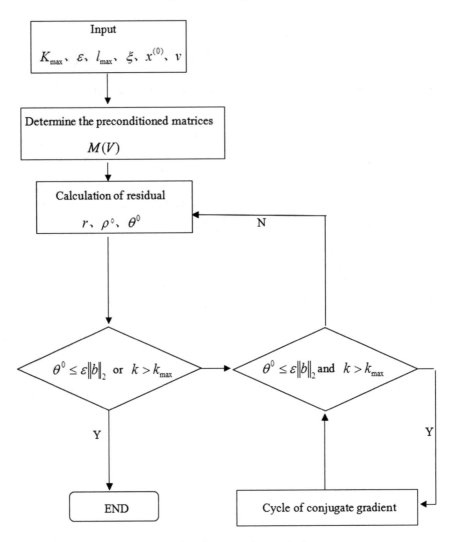

Fig. 2.3 Flowchart for preconditioned conjugate gradient method

converges quickly. The calculation process is shown in Fig. 2.3, where k_{max} is the maximum number of iterations for the outer loop, ε is the termination condition for the iterative regularized conjugate gradient method, l_{max} is the maximum number of iterations for the inner loop, ξ is the termination conditions for the inner conjugate gradient iteration, v is a predetermined regularization parameter.

$$M(v)^{-1}A(v)x = M(v)^{-1}(vx^k + F), \qquad (2.26)$$

where x^k is the value of the kth iteration of x, $x^{(0)}$ is the initial value that is selected to be a unit vector. $A(v) = vI + A^TA$.

The selection of regularization parameter v is very important. Regularization parameters $v(\delta)$ is used to balance the trade off between the accuracy and stability of the solution. Zhdanov et al. (1988, 1993, 1997) put forward the adaptive algorithm for diminishing the regularization factor. By comparing with the L-curve method, they found that results of the adaptive algorithm are not inferior to L-curve method. Since L-curve method needs to determine the optimal regularization factor by multiple inversions, the computation cost is duplicated. In the adaptive algorithm only the regularization factor involved in the present inversion needs to be determined, so it can greatly reduce the computation time. According to the deviation theory, Wang (2003, 2007) proposed a repeat selection conjugate gradient method (RSCG) to calculate the optimal regularization factor. Compared with conjugate gradients for normal equations with minimization of the residual norm (CGNR) and conjugate gradients for normal equations with minimization of the error norm (CGNE), RSCG method is more stable.

According to the RSCG method proposed by Wang (2003, 2007), the optimal regularization parameters are selected based on part of prior information. The factor v gradually change, the initial value v_0 is the ratio of the data fitting functional and the stability functional. The initial value of the regularization parameter can be determined based on large amount of simulations. Empirically, we can take v_0 to be 0.00005 as the original iteration value. In the subsequent iteration, if the data fitting residual gets gradually smaller with iteration, the regularization factor can keep constant, otherwise, it is chosen after the following equation

$$v = v_0\xi^k, \quad k = 0, 1\ldots, \tag{2.27}$$

where $\xi > 1$ is an empirical coefficient, k denotes the kth iteration in the iterative process of the preconditioned regularized conjugate gradient (PRCG).

After completing the integral discretization and determining the regularization parameter, Eq. (2.13) is solved by the preconditioned regularization conjugate gradient method. Assuming that the theoretical wave field is $u(x, y, z, \tau) = 1$, the equation becomes a simple integral, and we obtain the function value $f(x, y, z, t) = 1/\sqrt{\pi t}$. The regularization parameters, shown in Table 2.10, are regularization parameters suitable for time interval [0.000078, 0.006280]. Using PRCG method, the inverse transformation results are shown in Fig. 2.4. The maximum error is less than 2 %, illustrating that the method presented in the paper meets our needs.

Table 2.10 Estimation of regularization parameters

Regular parameters (v)	Number of iterations	Mean square error (%)
0.0011285	308,226	2.006

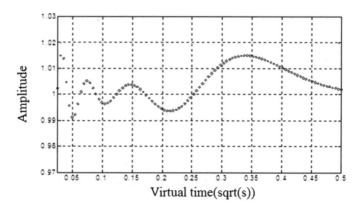

Fig. 2.4 Inverse transformation of wave field

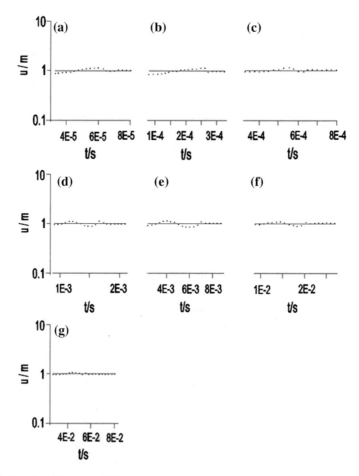

Fig. 2.5 Results of the wave field inversion

In order to validate the accuracy of the numerical scheme, the acquired integral coefficients of each time window are inversed into the wave field value. The numerical results of wave field value are compared with its theoretical value. Results of the inverse wave field and the theoretical wave field value corresponding to each time window are shown in Fig. 2.5. In Fig. 2.5, the dotted lines represent the inverse wave field value, while solid lines represent the theoretical wave field value $(u(\tau) = 1)$. The inversion results and level of obtained accuracy demonstrate that it is appropriate to separate the time range into seven time windows and use the coefficients in calculating the wave field values. It is also shown that relative errors for the seven time windows are 5.7, 6.7, 5.4, 6.0, 5.2, 5.7, and 4.5 %, respectively. In general, this error level is acceptable in practical engineering issues.

2.5   Correlation Stacking Method for Extracting Pseudo Wave Field

Due to the ill-posedness of wave field transformation equation, the wave field transformation matrix has large condition number which makes the numerical process very unstable. Preconditioned conjugate gradient regularization given in Sect. 2.4 may reduce the resolution to a certain extent. In this section, a new stable wave field transformation algorithm, namely the correlation stacking wave field transformation algorithm is addressed. Compared to the original regularized wave field transformation, the numerical experiments show that this method has good numerical stability with better wave field extraction.

Li et al. (2005a, b, c, d) proposed that good results can be achieved using segmentation method to reduce the condition number of the inverse transform matrix and the segmented wave field was transformed. However, the transformed wave field in each segment is not continuous in the vicinity of the endpoints. This leads problem of the convergence of the transformed wave field. We use the cor-relation stacking method to solve this problem. First, we extract pseudo-seismic wave field of whole time-domain TEM data with preconditioned regularized con-jugate gradient method, then pseudo-seismic wave field data of each time quantum computed based on same method. Finally, stack the pseudo-seismic wave field data according to their correlation, and regard the stacking result as the final extracting result. The time window chosen is shown in Fig. 2.6.

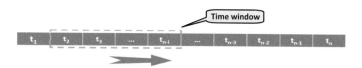

Fig. 2.6 Time window for wave field transformation

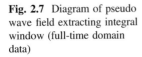

Fig. 2.7 Diagram of pseudo wave field extracting integral window (full-time domain data)

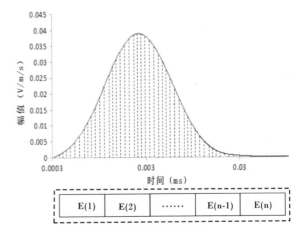

For correlation stacking method, there are six steps, which are as follows:

(1) Using recondition regularized conjugate gradient method on full time-domain data (Fig. 2.7) to extract pseudo wave field, the result is U_{all};

(2) Choose a time window and use preconditioned regularized conjugate gradient method on the data in this time window (Fig. 2.8) to extract pseudo wave field, the result is U_1,

(3) Move the chosen time window for a time-channel unit and use preconditioned regularized conjugate gradient method on the data in this time window to extract pseudo wave field, the result is U_2, (Fig. 2.9)

(4) Moving the whole time window for a time-channel unit again, and repeat step (3) until moving to the last time window, then pseudo wave field $U_3, U_4, \ldots, U_{n-m+1}$ extracted from each time window can be obtained;

Fig. 2.8 Pseudo wave field extracting integral window diagram of the first window

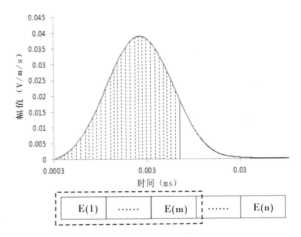

Fig. 2.9 Pseudo wave field extracting integral window diagram of the second window

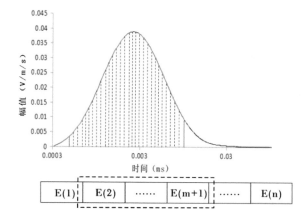

(5) Conduct correlation analysis on $U_n (n = 1, 2, \ldots, n - m + 1)$ and U_{all}, if their correlation is bigger than a threshold value α, retain U_n, or abandon U_n. The correlation between U_n and U_{all} is defined as,

$$\eta = \frac{|U_n \cdot U_{all}|}{|U_n| \cdot |U_{all}|}$$

(6) Stack all the retained U_n with U_{all}, and regard the stacking result as the finial result of pseudo wave field extracting.

2.6 Effectiveness Test of the Algorithm

To verify the validity and precision of the transformation algorithm introduced in last section, we use an analytical wave field solution and corresponding diffusion field to conduct wave field inverse transformation. We select the model as shown in Fig. 2.10.

An infinite line transmitter is located in a homogeneous whole space; the line source is located in the y-direction at $x = z = 0$ with a unite current. The current is shut off at zero time, then the transmitting current source can be expressed as

$$S = \mu \delta(x) \delta(z) \delta(t - 0^+), \tag{2.28}$$

Using the wave field transformation (2.6), the corresponding expression for the wave field source is

$$F = \mu \delta(x) \delta(z) \delta(q - 0^+). \tag{2.29}$$

Fig. 2.10 A whole space
model

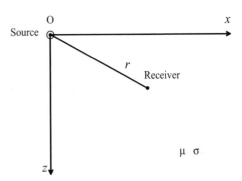

The field responses corresponding respectively to the source F and S in the whole space are

$$E = \frac{\mu}{4\pi} \frac{e^{-\mu\sigma\rho^2/4t}}{t}, \tag{2.30}$$

$$U = \frac{\mu}{2\pi} \frac{H(q - \sqrt{\mu\sigma}\rho)}{\sqrt{q^2 - \mu\sigma\rho^2}}, \tag{2.31}$$

where $\rho = \sqrt{x^2 + z^2}$, H is the unit step function.

The time window used is relatively small, so that the condition number of inverse wave field transformation equation is not big. Using the singular value decomposition will not exacerbate the degree of ill-posed equation, but instead makes the calculation faster.

Figure 2.11 shows the results of the wave field forward and inverse transformation using the singular value (SVD) decomposition. In Fig. 2.11a, analytic curve is calculated though forward Eq. (2.30), while numeric curve is calculated by using Eqs. (2.13) and (2.31) based on the method raised in 2.5 chapter. Similarly, in Fig. 2.11b, analytic curve is calculated though inverse Eq. (2.31), while numeric curve is calculated by using Eqs. (2.13) and (2.30) based on the method raised in 2.5 of this chapter. It is demonstrated that two curves agree with each other, which means that correlation stacking wave field transformation method can enhance the resolution relatively.

We used the method listed in Fig. 2.3 to process the data from each time window in Fig. 2.12, it is shown clear that the converted curves of every stack window are not smooth. After correlation stacking every curves of Fig. 2.12, the correlation stacking wave field transformation results is shown in Fig. 2.13. It is obvious that the correlation stacking wave field transformation has a very good application; the jumping values in the converted wave field disappear, in good agreement with the analytical solution.

In order to understand the sensitivity of this method to error, we add respectively 5 and 10 % white noise to the time-domain data and conduct the inverse wave field transformation. The results are shown in Fig. 2.10. It is seen from the figure that

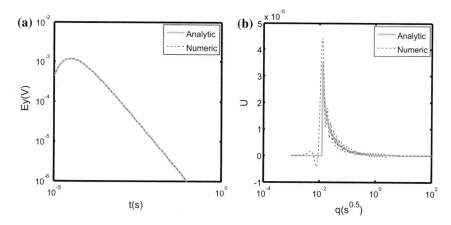

Fig. 2.11 Wave field transformation by using SVD algorithm. **a** Forward transformation. **b** Inverse transformation

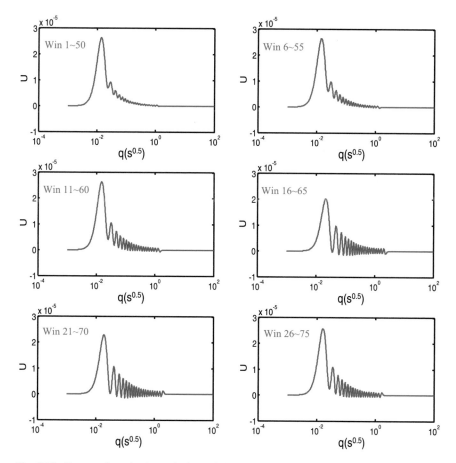

Fig. 2.12 Converted results for each time window

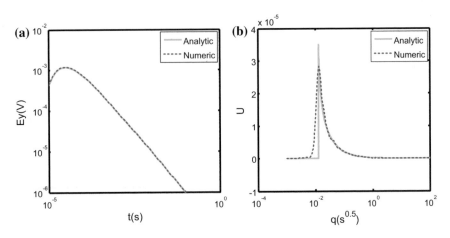

Fig. 2.13 Correlation stacking wave field transformation results. **a** Forward transform. **b** Inverse transform

after adding the white noise, the results of wave field transformation can still maintain the original waveform, indicating that the irrelevance between noises and signal, the superposition of correlation stacking method can well suppress the noise in time-domain (Fig. 2.14).

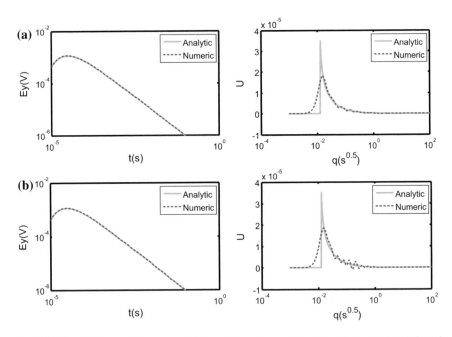

Fig. 2.14 Correlation stacking wave field transformation results with **a** 5 % noise added; and **b** 10 % noise added

References

Gershenson M (1993) Simple interpretation of time domain electromagnetic sounding using similarities between wave and diffusion propagation. Geophysics 62(3):763–774

Lee KH, Liu G, Morrison HF (1989) A new approach to modeling the electromagnetic response of conductive media. Geophysics 54(6):1180–1192

Li X, Xue GQ, Song JP, Guo WB, Wu JJ (2005a) The optimization of transient electromagnetic field-wave field conversion. Chin J Geophys 48(5):1185–1190 (in Chinese)

Li X, Song JP, Ma Y et al (2005b) The abstract of TEM signal based on the wavelet analysis. Coal Geol Explor 33(2):72–75

Li X, Xue GQ, Song JP et al (2005c) An optimize method for transient electromagnetic field-wave field conversion. Chin J Geophys 48(5):1185–1190

Li Xiu, Xue Guo-qiang, Song Jian-ping et al (2005d) Application of the adaptive shrinkage genetic algorithm in the feasible region to TEM conductive thin layer inversion. Appl Geophys 2(4):204–210

Maxwell AM (1996) Joint inversion of TEM and distorted MT soundings: effective practical considerations. Geophysics 61:P56–P65

Sternberg BK, Washburne JC, Anderson RG (1985) Investigation of MT static shift correction methods. SEG Tech Prog Expanded Abstr 4:264–267

Wang HJ (2003) Study on 1-D forward and inversion of transient electromagnetic. Doctoral Dissertation Submitted to Xi'an Jiao Tong University, Xi'an

Wang YF (2007) Calculation method and application of inversion problem. Higher Education Press, China

Zhdanov MS, Booker JR (1993) Underground imaging by electromagnetic migration. In: 63rd Ann Internat Mtg Expl Geophys, Expanded Abstr, pp 355–357

Zhdanov MS, Matusevich VU, Frenkel MA (1988) Seismic and electromagnetic migration. Nauka (in Russian)

Zhdanov MS, Portniaguine O (1997) Time-domain electromagnetic migration in the solution of inverse problems. Geophys J Int 131:293–309

Chapter 3
Property of TEM Pseudo Wave Field

The diffusion equation TEM field satisfies mainly describes the diffusion characteristics of eddy current, which may cause strong volume effect. Thus, the migration and imaging method based on the diffusion equation generally has poor capability to distinguish electrical interfaces. The purpose of wave field transform is to extract information connected with propagation from EM responses and suppress or remove that connected with dispersion and attenuation. Numerical experiments indicate that the pseudo wave field obtained by the wave field transformation not only satisfies wave equation, but it also has property of propagation, reflection, and transmission, similar to seismic wavelet.

In this chapter, we use wave field transformation to analyze the responses of two- and three-layer models, give the virtual time and wavelet width, and summarize the characteristics of virtual wave field on reflection, refraction, and absorption and attenuation. This is important for studying the minimum resolution of pseudo wave field.

3.1 Two-Layer Model Analysis

Using special wavelet $u(x, y, z, \tau) = 1$, we can determine the reasonable range of regularization parameters. After obtaining these parameters, we can build the virtual wave field from the Gaussian pulses as wavelets and simulate the earth model.

3.1.1 Wave Field with Single Positive Peak

Synthesized seismogram with Gaussian pulses as wavelets is shown in Fig. 3.1a. Figure 3.1b shows the decay curve of pseudo wave field, Fig. 3.1c shows the

© Science Press and Springer Nature Singapore Pte Ltd. 2017

X. Li et al., *Migration Imaging of the Transient Electromagnetic Method*,

DOI 10.1007/978-981-10-2708-6_3

integral function of superimposed wavelets. The decay curve has the same shape as that of D-type model, so we assume single positive peak seismogram is wave field function of D-type earth model. Through analysis we know that the peak time of the impulse is related to electrical interfaces, the pulse width is related to resistivity contrast, while the sign is related to the resistivity contrasts. We will proceed with wave field transformation after choosing the optimal regularization parameters according to the discussion in Chap. 2.

Let us give a series of single peak seismogram. We fix the wavelet width W_1 and amplitude A_1, while changing the peak time T_1 in the range of 0.02–0.1 $\sqrt{s}$. We calculate the model responses for the flowing parameters: the sampling time: 10 μs–10 ms, the virtual time: 0.005–0.5 $\sqrt{s}$, the sampling points is $M = N = 101$, the optimal regularization parameter is 0.003.

Figure 3.2 shows the inversion results. The solid lines are synthesized seismograms $U'(\tau)$, while the circles and dashed lines are transformation results $U(\tau)$.

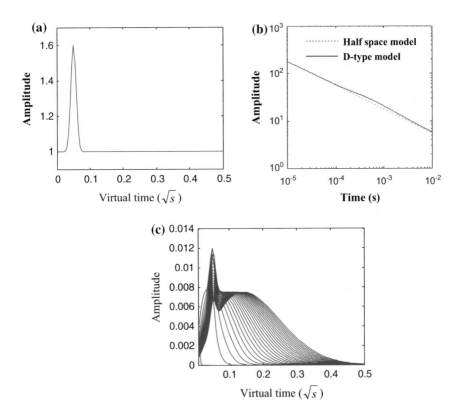

Fig. 3.1 Synthesized seismogram for a D-type earth model with Gaussian pulses as wavelets. **a** Pseudo seismogram; **b** decay curves of equivalent model; **c** kernel function for different sampling time

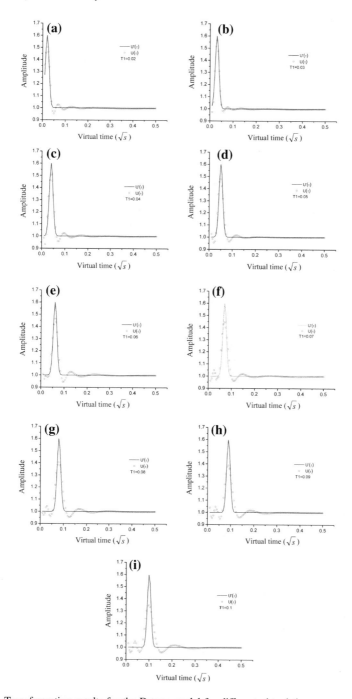

Fig. 3.2 Transformation results for the D-type model for different virtual time

Table 3.1 Iterations and mean square deviation for different virtual time

No.	Peak time (T_1)	Wavelet amplitude (A_1)	Wavelet width (W_1)	Iteration numbers	Mean square error (%)
a	0.02	0.6	0.04	66,941	0.83
b	0.03	0.6	0.04	66,203	0.54
c	0.04	0.6	0.04	60,065	0.96
d	0.05	0.6	0.04	61,031	1.22
e	0.06	0.6	0.04	63,133	1.38
f	0.07	0.6	0.04	63,692	1.7
g	0.08	0.6	0.04	60,975	1.91
h	0.09	0.6	0.04	57,203	2.03
i	0.1	0.6	0.04	61,172	2.41

Fig. 3.3 Mean square error versus peak time

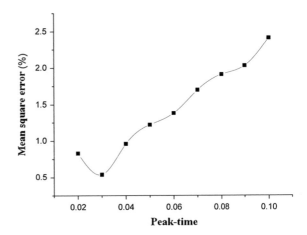

    From Fig. 3.2, we can see that with the delay of the peak time, the wavelet amplitude reduces and the width slightly broadens. Table 3.1 shows the calculation results. From the table, the mean square error increases gradually with virtual time, as shown in Fig. 3.3. This is consistent with the propagation characteristic of EM waves in a dispersive medium, which reduces the resolution of TEM method.

    Figure 3.4 shows the transformation results for fixed peak time but with variable wavelet widths. The circles are for transformation results, while the solid lines are for synthesized seismogram. The sampling interval and virtual time interval stay unchanged.

    Table 3.2 shows the analysis results of wave field transformation. From Fig. 3.4, we can see that at the same depth the difference between the inverted amplitude and the model amplitude and the mean square error becomes bigger with narrowing wavelet widths. When wavelet widths become narrow enough (resistivity changes very slightly), the transformed wavelets cannot be distinguished. Figure 3.5 shows the relationship between the transformation error and the wavelet widths. From the

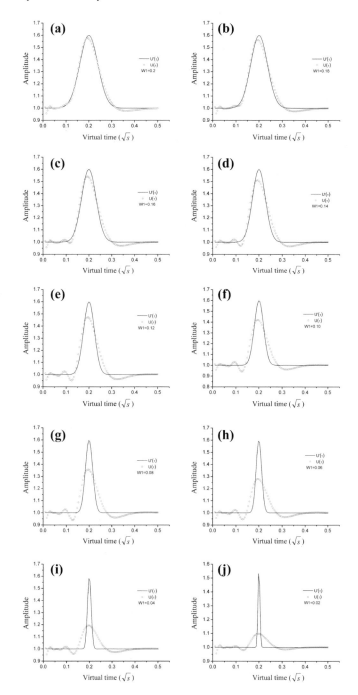

Fig. 3.4 Transformation results for the D-type model with fixed peak time but different wavelet widths

Table 3.2 Inversion mean square error for a D-type earth model with fixed depth

No.	Peak time (T_1)	Wavelet amplitude (A_1)	Wavelet width (W_1)	Iterations	Mean square error (%)
(a)	0.2	0.6	0.2	62,863	1.49
(b)	0.2	0.6	0.18	62,373	1.78
(c)	0.2	0.6	0.16	60,892	2.12
(d)	0.2	0.6	0.14	61,301	2.57
(e)	0.2	0.6	0.12	56,076	3.09
(f)	0.2	0.6	0.10	58,696	3.58
(g)	0.2	0.6	0.08	61,728	3.95
(h)	0.2	0.6	0.06	61,882	4.02
(i)	0.2	0.6	0.04	60,665	3.53
(j)	0.2	0.6	0.02	62,295	2.34

Fig. 3.5 Transformation error versus wavelet width

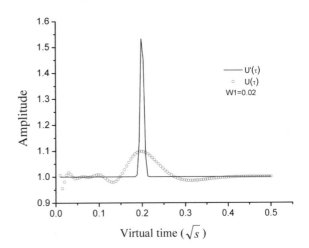

figure, we can see that when wavelet width is very small, the mean square error becomes small, however, the maximum is relative small compare with minimum value in fitting curve (for example, red color curve in Fig. 3.4g), resulting in poor resolution for narrow wavelet width curve.

3.1.2   Wave Field with Single Negative Peak

Similar to the wave field analysis with single negative peak, we analyze in this section the single negative peak model. The synthesized seismogram with Gauss pulse as wavelets is shown in Fig. 3.6a. Figure 3.6b shows the decay curve of pseudo wave field, while Fig. 3.6c shows the integral function of superimposed

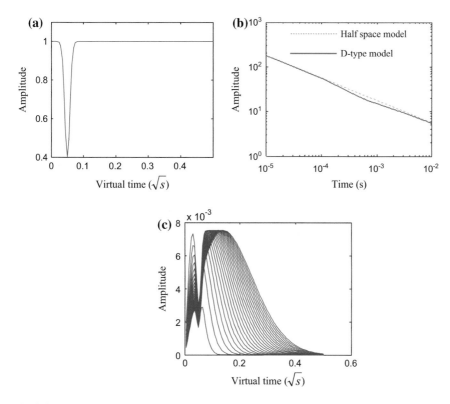

Fig. 3.6 Synthesized seismogram for a G-type earth model with Gaussian pulses as wavelets. **a** pseudo seismogram; **b** decay curves of equivalent model; **c** kernel function for different sampling time

wavelets. The shape of decay curve is the same as that of a G-type earth model. Thus, we assume a single negative peak seismogram to be a wave field function of G-type model and proceed with the inverse transformation as above. Figure 3.7 shows the transformation results with fixed wavelet width and amplitude, but with different peak time. The inversion uses the previous preconditioned regularization conjugate gradient method.

From Fig. 3.7, we can see that with delaying peak time, the amplitudes of wavelets decrease but the width increases slightly. This is consistent with the propagation characteristic of EM wave in a dispersive medium. Table 3.3 shows the error of wave field transformation. Figure 3.8 shows the transforming error versus peak time, from which we can see that the transformation error increases with peak time. These transformation results indicate that TEM has high resolution to shallow targets. With increasing depth, the high-frequency signal decays very heavily, resulting in poor resolution.

Figure 3.9 shows the transformation results for fixed peak time and variable wavelet width.

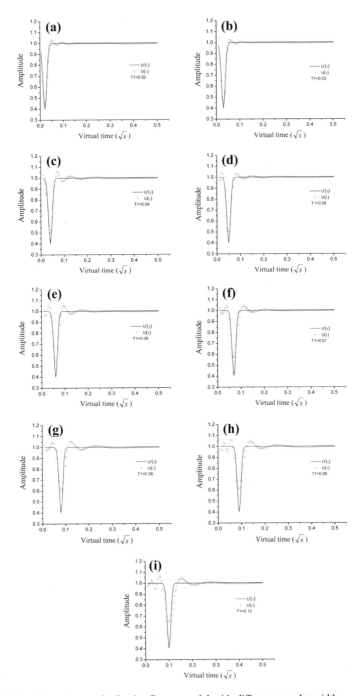

Fig. 3.7 Transformation results for the G-type model with different wavelet widths

Table 3.3 Transformation error for a G-type model with same resistivity

No.	Peak time (T_1)	Wavelet amplitude (A_1)	Wavelet width (W_1)	Iterations	Mean square error (%)
a	0.02	−0.6	0.04	59,407	0.59
b	0.03	−0.6	0.04	59,432	0.63
c	0.04	−0.6	0.04	63,599	1.07
d	0.05	−0.6	0.04	63,980	1.13
e	0.06	−0.6	0.04	60,526	1.38
f	0.07	−0.6	0.04	60,495	1.67
g	0.08	−0.6	0.04	64,805	1.9
h	0.09	−0.6	0.04	62,486	2.1
i	0.1	−0.6	0.04	61,505	2.51

Fig. 3.8 Transformation error versus peak time

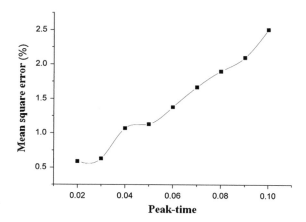

From Fig. 3.9, we can see that for the same peak time, the resolution increases with broadening wavelet widths. Refer to Table 3.4 and Fig. 3.10, except for the first two points, the transformation error decreases with increasing wavelet width. When wavelet width is 0.04, the transformation error is small and the virtual wavelet amplitude is one order larger than the amplitude of the maximum side lobe. When wavelet width is 0.02, the transformation error is small, but the virtual wavelet amplitude is only a few times of the amplitude of the maximum side lobe. Thus, if the wavelet width is too small, the pseudo wave field cannot well resolve the electrical interfaces.

Analysis of two-layer models shows that the wave field transformation for D- and G-type earth models are similar. At the same depth, the wavelet amplitude reduces with decreasing resistivity contrasts, the underground electrical interfaces cannot be distinguished when wavelet width becomes small enough. For the same resistivity contrasts, the amplitudes of pseudo wave field wavelets attenuate with the peak time, while the wavelet width became broad.

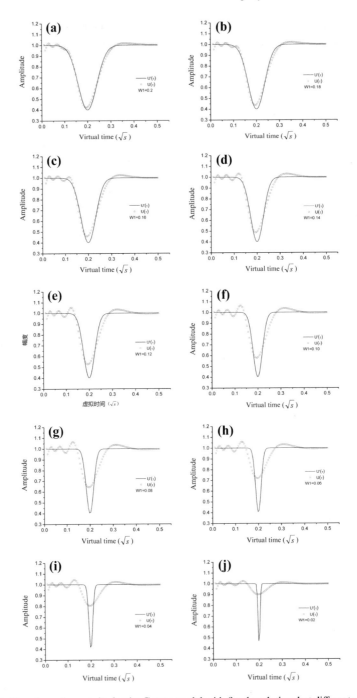

Fig. 3.9 Transformation results for the G-type model with fixed peak time but different resistivity

Table 3.4 Transformation error for a G-type earth model with constant first-layer parameters

No.	Peak time (T_1)	Wavelet amplitude (A_1)	Wavelet width (W_1)	Iterations	Mean square error (%)
a	0.2	−0.6	0.2	62,519	1.44
b	0.2	−0.6	0.18	63,641	1.72
c	0.2	−0.6	0.16	60,894	2.1
d	0.2	−0.6	0.14	63,855	2.59
e	0.2	−0.6	0.12	64,006	3.14
f	0.2	−0.6	0.1	62,725	3.67
g	0.2	−0.6	0.08	62,404	4.06
h	0.2	−0.6	0.06	61,728	4.15
i	0.2	−0.6	0.04	62,643	3.65
j	0.2	−0.6	0.02	64,018	2.44

Fig. 3.10 Transformation error for the G-type model versus the wavelet width

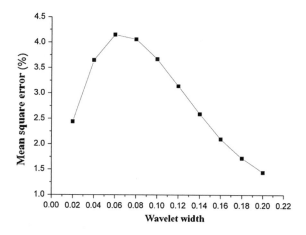

3.2   Three-Layer Model Analysis

Electromagnetic wave propagates and attenuates fast in a conductive medium, so that it is difficult to identify the interfaces of a multilayer earth. To analyze the resolution of wave field transform for multi-layer models, we design models with two wavelets and analyze the transformation results.

3.2.1   Q-Type Model with Double Positive Peaks

The synthesis seismogram (Q-type model) is shown in Fig. 3.11a. The attenuation curve of pseudo wave field after transformation is shown as solid lines in Fig. 3.11b. For comparison, the dotted lines give the decay curve for a

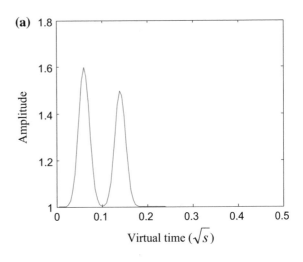

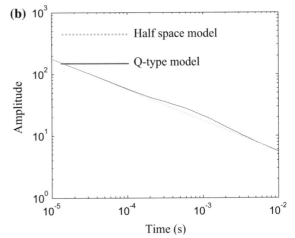

homogeneous half-space without any reflection interfaces. The resistivity reduces from layer to layer, resulting in the gradually slowing attenuation in the decay curve.

This is similar to the EM diffusion in the model medium, thus we can take the synthesis seismogram as that of the virtual wave field for Q-type earth model.

We carry out the inversions using the above introduced method. For easy of analysis, we first choose the width of the wavelet, fix the interval between the two peak times, while changing the first peak time T_1. The transformation results are shown in Fig. 3.12a–c. Then, we fix the first peak time T_1, but reduce the second peak time T_2. The transformation results are shown in Fig. 3.12d–f.

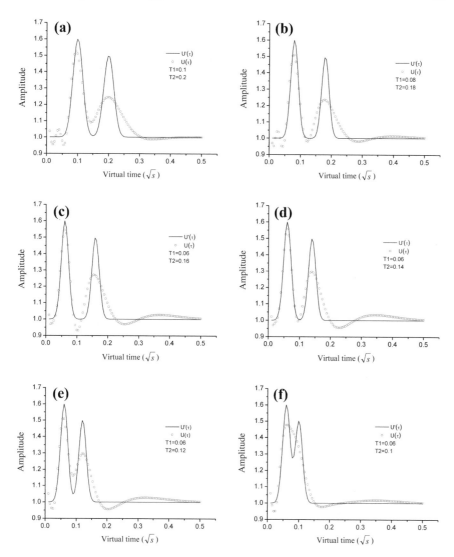

Fig. 3.12 Transformation results for the Q-type model

From the transformation results for the Q-type model in Fig. 3.12, we can see that for the model of multiple layers, with delaying peak time, the wavelet amplitudes attenuate gradually. The second wavelet seriously attenuates. As the interval between two peak times reduces, the two wavelets gradually overlap, resulting in poor resolution to the thin layer. Table 3.5 shows the wave field transformation error. When wavelet interval is 0.04, the two wavelets overlap each other, so that thin layer cannot be distinguished.

Table 3.5 Transformation error for a Q-type earth model

No.	First peak time (T_1)	First wavelet width (W_1)	First wavelet amplitude (A_1)	Second peak time (T_2)	Second wavelet width (W_2)	Second wavelet amplitude (A_2)	Iterations	Mean square error $(\%)$
a	0.1	0.06	0.6	0.2	0.06	0.5	1,269,213	4.7582
b	0.08	0.06	0.6	0.18	0.06	0.5	1,277,407	3.8505
c	0.06	0.06	0.6	0.16	0.06	0.5	1,271,561	3.7838
d	0.06	0.06	0.6	0.14	0.06	0.5	1,272,517	3.4252
e	0.06	0.06	0.6	0.12	0.06	0.5	1,272,594	3.7434
f	0.06	0.06	0.6	0.1	0.06	0.5	1,267,121	2.8502

3.2.2   H-Type Model with Positive–Negative Peaks

Figure 3.13a shows the synthesized seismogram, Fig. 3.13b shows the decay curve of corresponding time-domain diffusion field. Due to the fact that the resistivity of a H-type earth decreases first and then increases, in the decay curve the model responses decay initially slow and then fast, coincident with the TEM propagation characteristic. Thus, the synthesized seismogram reflects the H-type virtual wave field. Using the previous method, we carry out the inverse transformation and compare the results with the known wave field, the results are displayed in Fig. 3.14.

To carry out the wave field transformation, we first fix the wavelet width and the first peak-time T_1 and shift the second peak time T_2. The transformation results are shown in Fig. 3.14a–e. Then, we fix the interval of two peak times and shift the second peak time (T_2). The transformation results are shown in Fig. 3.14e–g.

From Fig. 3.14a–e, we can see that since the first peak time is fixed, the first transformed wavelet fits the theoretical one well. As the interval between the two peak times increases, the amplitude of the second transformed wavelet attenuates gradually, the transformation error increases. From Fig. 3.14e–g, we can see that as the first peak time increases, the amplitude of transformed wavelet, especially the second one, attenuates. The transformation error increases as well (Table 3.6).

3.2.3   K-Type Model with Negative–Positive Peaks

The synthesized seismogram for this model is shown in Fig. 3.15a, while the corresponding time-domain decay curve is shown in Fig. 3.15b. The change of decay curve is coincident with the K-type of earth model. The EM signal decays

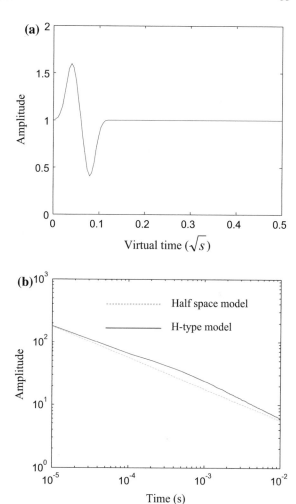

Fig. 3.13 Synthesized seismogram for the H-type model with Gaussian pulses as wavelet. **a** pseudo seismogram; **b** decay curves of equivalent model

fast at the early time but slowly at the later time. Thus, the synthesized seismogram with negative–positive peaks is taken as that of the K-type model. In Fig. 3.15b, the solid line corresponds to the K-type model, while dotted line corresponds to a homogeneous half-space.

Based on the method addressed in Chap. 2, we run the wave field transformation for the K-type model. We first fix the interval of two peak times and shift the first peak time, the transformation results are shown in Fig. 3.16. From the figure, we can see that as the first peak time shift to the later time, the amplitude of transformed wavelets decreases gradually, especially the second wavelet decays faster than the first wavelet. The transformation error gradually increases (Table 3.7).

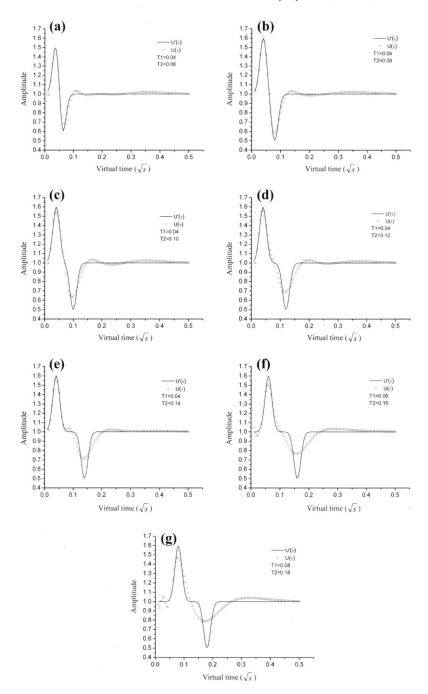

Fig. 3.14 Transformation results for the H-type model

Table 3.6 Transformation error for an H-type earth model

No.	First peak time (T_1)	First wavelet width (W_1)	First wavelet amplitude (A_1)	Second peak time (T_2)	Second wavelet width (W_2)	Second wavelet amplitude (A_2)	Iterations	Mean square error (%)
a	0.04	0.06	0.6	0.06	0.06	−0.5	73,491	1.3743
b	0.04	0.06	0.6	0.08	0.06	−0.5	71,107	1.7725
c	0.04	0.06	0.6	0.10	0.06	−0.5	1,276,072	3.032
d	0.04	0.06	0.6	0.12	0.06	−0.5	1,273,393	2.9119
e	0.04	0.06	0.6	0.14	0.06	−0.5	1,275,730	2.8213
f	0.06	0.06	0.6	0.16	0.06	−0.5	1,269,651	4.3382
g	0.08	0.06	0.6	0.18	0.06	−0.5	1,274,553	5.3975

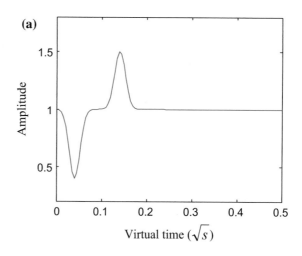

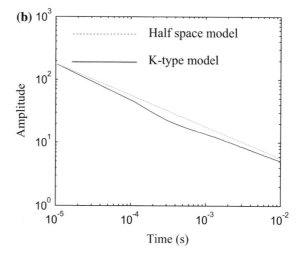

Fig. 3.15 Synthesized seismogram for a K-type earth model with Gaussian pulses as wavelets. **a** pseudo seismogram; **b** decay curves of equivalent model

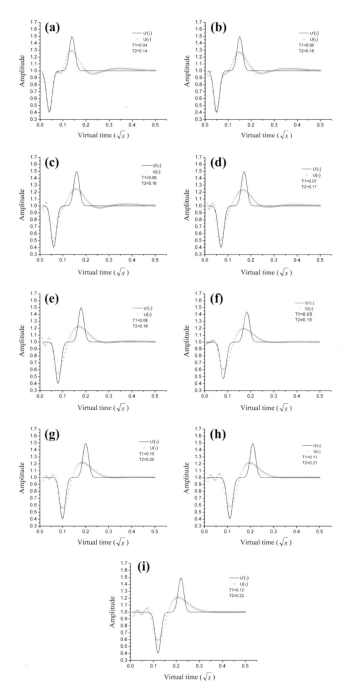

Fig. 3.16 Transformation results for the K-type earth model

Table 3.7 Transformation error for the K-type earth model

No.	First peak time (T_1)	First wavelet width (W_1)	First wavelet amplitude (A_1)	Second peak time (T_2)	Second wavelet width (W_2)	Second wavelet amplitude (A_2)	Iterations	Mean square error $(\%)$
a	0.04	0.06	−0.6	0.14	0.06	0.5	1,272,259	3.373
b	0.05	0.06	−0.6	0.15	0.06	0.5	1,273,564	3.9016
c	0.06	0.06	−0.6	0.16	0.06	0.5	1,262,827	4.2297
d	0.07	0.06	−0.6	0.17	0.06	0.5	1,270,457	4.284
e	0.08	0.06	−0.6	0.18	0.06	0.5	1,270,215	4.6335
f	0.09	0.06	−0.6	0.19	0.06	0.5	1,271,489	3.8511
g	0.10	0.06	−0.6	0.20	0.06	0.5	1,276,221	4.7129
h	0.11	0.06	−0.6	0.21	0.06	0.5	1,267,150	4.9841
i	0.12	0.06	−0.6	0.22	0.06	0.5	1,273,142	5.0905

3.2.4 A-Type Model with Double Negative Peaks

The synthesized seismogram for this model is shown in Fig. 3.17a, while the corresponding time-domain decay curve is shown in Fig. 3.17b. The change of decay curve is coincident with the A-type of earth model. The resistivity in the earth increases with depth, resulting in the EM signal decaying faster and faster with time. Thus, the decay curve is coincident with the synthesized seismogram for an A-type earth model. In Fig. 3.17b, the solid line corresponds to the A-type model, while dotted line corresponds to a homogeneous half-space.

We use the same algorithm to run the wave field transformation for the A-type model. We fix the wavelet width and first peak time and shift the second peak time gradually. The transforming results are shown in Fig. 3.18, while the errors are given in Table 3.8. From Fig. 3.18, we can see that since the first peak time is fixed, the amplitude of first transformed wavelet does not change. As the second peak time is shifted toward early time, the amplitude of the second transformed wavelet increases. When the interval between two transformed wavelets is $0.04s^{1/2}$, the two transformed wavelets overlap each other and the interfaces can no longer be distinguished.

3.3 Wave Field Characteristics of Time-Domain EM Responses with Noise

Field data contain all kinds of noise, so it is important to analyze pseudo wave field property of time-domain EM response with noise. We add Gauss white noise to time-domain response with S/N ratio of 100:1, 50:1, 30:1 and 20:1. The time-

Fig. 3.17 Synthesized
seismogram for an A-type
earth model with Gaussian
pulses as wavelets. **a** pseudo
seismogram; **b** decay curves
of equivalent model

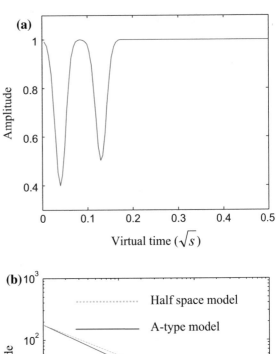

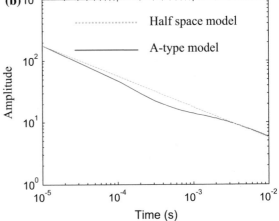

domain responses with noise and the corresponding transformation results are
shown in Figs. 3.19 and 3.20.

From Fig. 3.19d, we can see that the time-domain response is not smooth
because of the noise. From Fig. 3.20, when S/N ratio becomes 50:1, the transformation result is similar to the theoretical one. When S/N ratios respectively 30:1
and 20:1, the transformation results have strong oscillation, resulting in great difference from the theoretical responses.

Let the virtual wave assume different waveforms of a single positive Gauss
pulse, a single negative pulse, double positive pulses, double negative pulses,

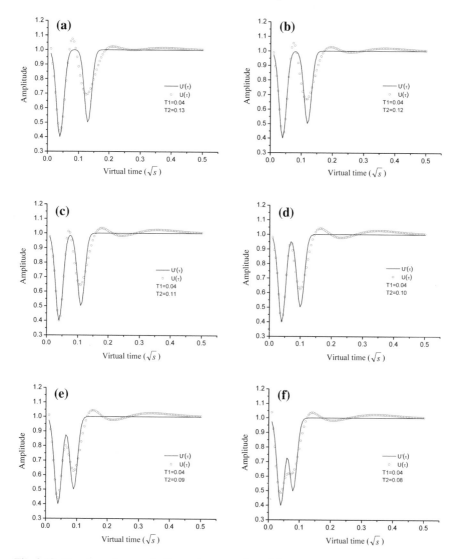

Fig. 3.18 Transformation results for an A-type earth model

positive–negative pulses, and negative–positive pulses, the transformation results
are displayed in Figs. 3.21, 3.22, 3.23, 3.24, 3.25, 3.26, 3.27, 3.28, 3.29, 3.30, 3.31
and 3.32.

From Figs. 3.22 and 3.24, we can see that when S/N ratio is 30:1, the wave field
transformation result is similar to the theoretical one, the single peak is accurately

Table 3.8 Transformation results for an A-type earth model

No.	First peak time (T_1)	First wavelet width (W_1)	First wavelet amplitude (A_1)	Second peak time (T_2)	Second wavelet width (W_2)	Second wavelet amplitude (A_2)	Iterations	Mean square error (%)
a	0.04	0.06	−0.6	0.13	0.06	−0.5	1,269,525	0.029219
b	0.04	0.06	−0.6	0.12	0.06	−0.5	1,276,592	0.026293
c	0.04	0.06	−0.6	0.11	0.06	−0.5	1,275,498	0.024571
d	0.04	0.06	−0.6	0.10	0.06	−0.5	1,272,992	0.023016
e	0.04	0.06	−0.6	0.09	0.06	−0.5	1,274,703	0.025678
f	0.04	0.06	−0.6	0.08	0.06	−0.5	1,269,852	0.025171

Fig. 3.19 Time-domain EM responses with added noise, $U(\tau) = 1$

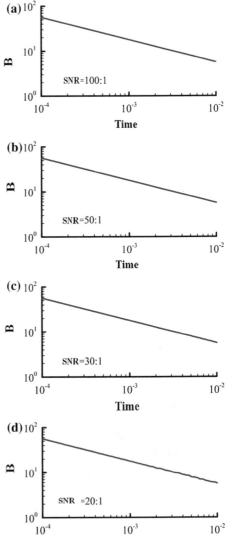

Fig. 3.20 Transforming result of time-domain response with noise when $U(\tau) = 1$

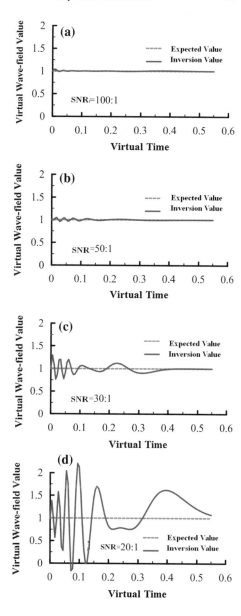

positioned. When noise–signal ratio is 20:1, the transformation result has a great difference with theoretical result with strong oscillations, the position of single peak cannot be distinguished. From Figs. 3.21 and 3.23, we can see that the time-domain

Fig. 3.21 Time-domain responses with noise for single positive pulse

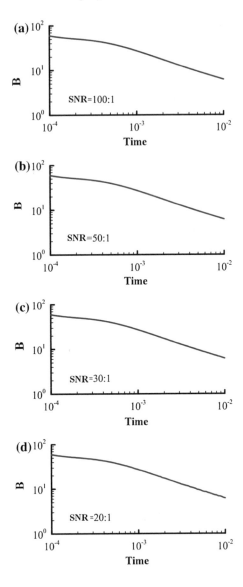

responses are very rough when the S/N ratio is 20:1, there exists serious interference. This indicates that the data smoothness is very important.

From Fig. 3.26, we can see that when the S/N ratio is 30:1, the first peak position of transformation result coincides with theoretical one, but the second peak position is not accurate. Since Gauss white noise is random, the results may not be the same

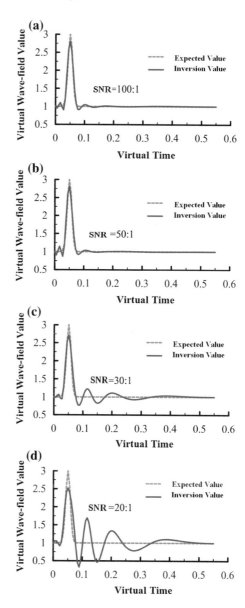

Fig. 3.22 Transformation results of noisy time-domain data for single positive pulse

for each transformation. From Fig. 3.28, we can see that when the S/N ratio is 30:1, the positions of two peaks from the transformation results coincide well with theoretical results. From Figs. 3.26 and 3.28, we can see that when the S/N ratio is 20:1, all transforming results have large differences from theoretical ones, making two peaks difficult to distinguish.

Fig. 3.23 Time-domain
responses with noise for
single negative pulse

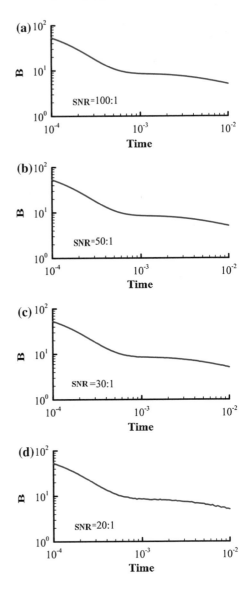

From Figs. 3.29 and 3.31, we can see that when the S/N ratio is 30:1, the curves for time-domain EM responses are not smooth. When the S/N ratio reduces to 20:1, the curves for time-domain responses have obvious burrs. From Figs. 3.30 and 3.32, we can see that when the S/N ratio is not less than 30:1, the transformation

Fig. 3.24 Transformation results of noisy time-domain data for single negative pulse

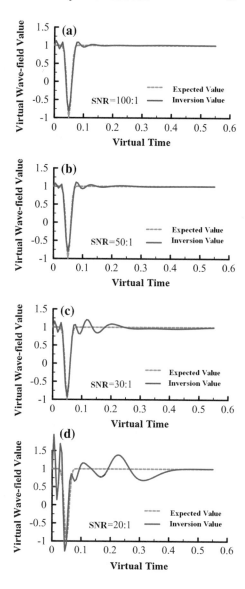

results coincide well with theoretical results. However, when the S/N ratio is 20:1, the peak position is not accurately determined. There exist several positive and negative peaks, resulting in false results.

Since for inverse wave field transformation, the condition number of coefficient matrix is very big, the solution of the equations system is not stable. The tiny

Fig. 3.25 Time-domain
responses with noise for
double positive pulses

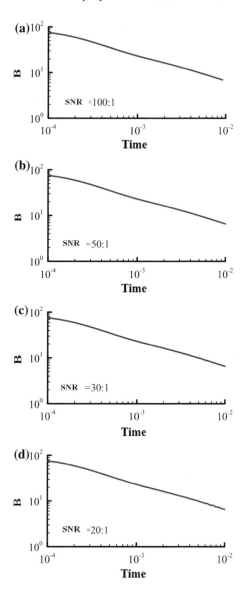

disturbance on the right side can result in big influence on the solution. Noise
experiments show that when the S/N ratio is not less than 30:1, the transformation
results well coincide with theoretical ones.

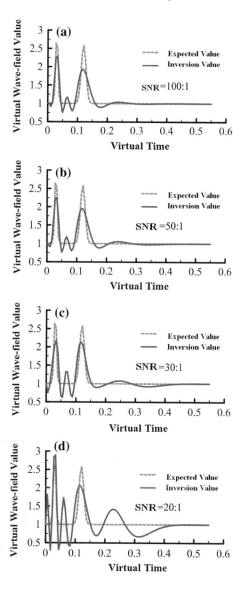

Fig. 3.26 Transformation results of noisy time-domain data for double positive pulses

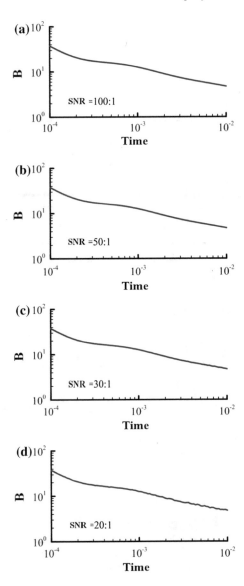

Fig. 3.27 Time-domain responses with noise for double negative pulses

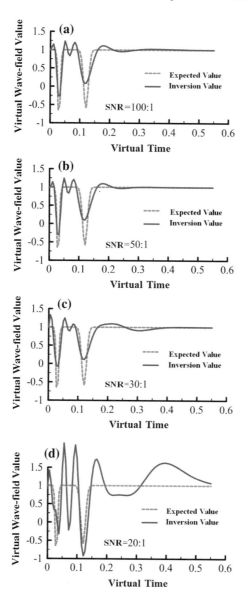

Fig. 3.28 Transformation results of noisy time-domain data for double negative pulses

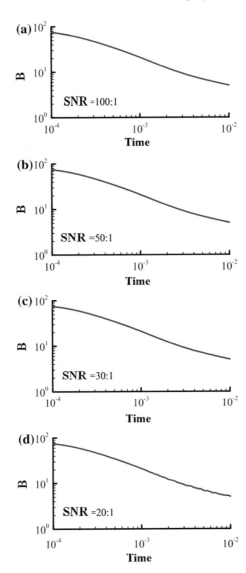

Fig. 3.29 Time-domain responses with noise for positive–negative pulses

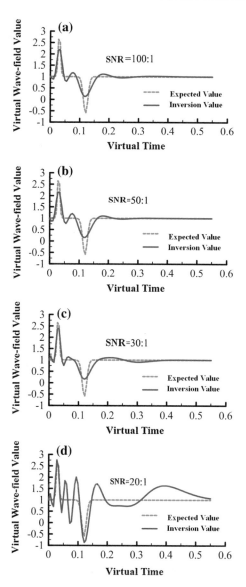

Fig. 3.30 Transformation results of noisy time-domain data for positive–negative pulses

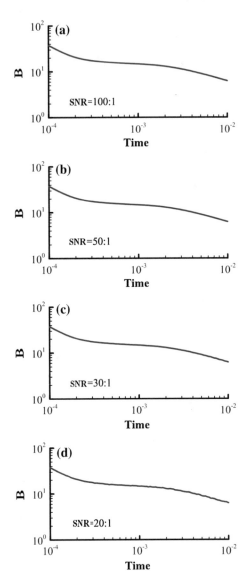

Fig. 3.31 Time-domain responses with noise for negative–positive pulses

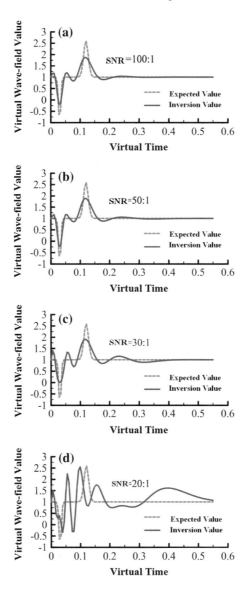

Fig. 3.32 Transformation results of noisy time-domain data for negative–positive pulses

In summary, the accuracy of inverse wave field transform rely heavily on the observation accuracy of TEM responses. Only if observation accuracy is high enough, can an inverse wave field transformation is meaningful.

Chapter 4
Synthetic Aperture Algorithms and Compression of Wavelet Width

4.1 Imaging Method Based on Synthetic Aperture

After putting forward the wave field transformation method, we have realized the transformation of the diffusive TEM field into a pseudo-seismic wave field. This creates the conditions for the realization of the synthetic aperture imaging of airborne transient EM method. The imaging method based on synthetic aperture technology for TEM takes advantage of the concept of synthetic aperture radar imaging, which uses the relative motion between the target and the real airborne aperture transmitting coil to synthetize via data processing a large equivalent aperture for the transmitting coil from the small-sized real antenna aperture, to improve the resolution and penetration capability. For airborne TEM method, the observation configuration is similar to airborne synthetic aperture radar (SAR), the imaging theory and technology of synthetic aperture radar can be used to realize the TEM synthetic aperture imaging based on the pseudo-seismic field transformation.

After the wave field transformation, the TEM diffusive field is converted into wave field. The data at each survey station is equivalent to the wave field of the self-excitation and self-receiving. Through analysis on numerical experimental results, people have determined that the TEM fields for multiple transmitting sources have the property of correlated superposition. Based on these TEM characteristics, we use the correlated superposition to carry out the synthetic aperture. A schematic diagram is shown in Fig. 4.1.

To run a correlated superposition for synthetic aperture, we first select a central point that is taken as ith point, the value of the wave field at this point can be expressed as $U(r_i, \tau)$, where r_i is the distance between the ith point to one point among $-N, \ldots, N$, τ is the relative time shift. We then choose the length of $2N + 1$

© Science Press and Springer Nature Singapore Pte Ltd. 2017
X. Li et al., *Migration Imaging of the Transient Electromagnetic Method*,
DOI 10.1007/978-981-10-2708-6_4

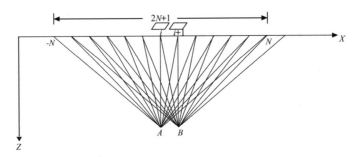

Fig. 4.1 Schematic diagram of synthetic aperture

survey points as the length for synthetic aperture, namely we select the N points respectively to the left and to the right of the ith point and calculate the correlation with the central point. The normalized cross-correlation coefficient is given by

$$\rho(r_i, \tau) = \frac{\sum_{j=1}^{m} U(r_i, t_j) U(r_{i+k}, t_j - \tau)}{\left\{ \sum_j [U(r_i, t_j)]^2 \sum_j [U(r_{i+k}, t_j)]^2 \right\}^{\frac{1}{2}}}, \tag{4.1}$$

where m is the time channels for each survey point.

The cross-correlation coefficient $\rho(r_i, \tau)$ represents the degree of correlation of two wave fields that is related to the relative time shift τ. By changing τ, we can find the optimal time delay denoted as τ for the maximum correlation coefficient $\rho(r_i, \tau)$. Thus, we can obtain $2N + 1$ maximum correlation coefficients $\rho^{\max}(r_k, r_k^m)$ and $2N + 1$ optimal delay time τ_k^m. Then, the maximum correlation coefficients obtained from the correlation calculation are recorded as the weighting coefficients, we can obtain the synthetic value at the central point by multiplying the wave field at each point with the corresponding coefficients and superimpose them to the central point. This is expressed as

$$U'(r_i, t_j) = \sum_{K=-N}^{N} \rho^{\max}(r_k, \tau_k^m) U(r_k, t_j - \tau_k^m) \quad (j = 1, 2, \cdots, m), \tag{4.2}$$

Repeating this process along a survey line, we can obtain the synthesis values points $i + 1$, $i + 2$, $i + 3$, etc.

Here, we only show an example by discussing the synthetic aperture algorithm for one-dimensional (1D) profile. This method can be extended to a two-dimensional (2D) synthetic aperture problem. Further research can even develop algorithms based on self-focusing.

4.2   Compression of Wavelet Width

Although the transformed wave field and the seismic wave field all satisfy the wave equation, however, due to the different physical background of two fields, there exist important differences between them. The former is the "reflex" wavelet associated with the inductive TEM attenuation, it is virtual, while the latter is the seismic wavelet existing objectively in an elastic medium. In addition, the propagation speed of the pseudo-seismic wave depends not only on the conductivity of the medium, but also on the conductivity of adjacent medium. This is very different from the real seismic wave. A serious problem caused by this essential difference is the wave dispersion effect, which will decrease the resolution of the obtained virtual wave field. This will be a great drawback in the TEM interpretation by using the pseudo-seismic method. Obviously, the wave dispersion effect will greatly influence the space resolution of TEM imaging and limit the TEM application in geological surveys. Therefore, it is necessary to explore the cause and how to control the wave dispersion effect.

In the following, we discuss the cause for the dispersion the virtual wave field. We will see that the wave dispersion is not produced by the energy loss while propagating through the media, but is caused by the increase of the distribution range of the Gauss distribution in the kernel function with the virtual time in the wave field transformation. Finally, we remove the wave dispersion through computing the deconvolution of the virtual wave filed data. The experiments with synthetic data confirm that the method can control the wave dispersion. Thus, it can enhance the TEM capability to distinguish the underground electrical structures.

4.2.1   The Phenomenon of Waveform Dispersion

In order to explain the waveform dispersion in wave field transformation, we selected three models (K-, A-, and Q-type), the geoelectric parameters for these models are as follows:

K-type model: $\rho_1 = 5 \, \Omega \cdot m$, $h_1 = 60 \, m$, $\rho_2 = 50 \, \Omega \cdot m$, $h_2 = 60 \, m$, $\rho_3 = 5 \, \Omega \cdot m$;
A-type   model: $\rho_1 = 5 \, \Omega \cdot m$, $h_1 = 60 \, m$, $\rho_2 = 50 \, \Omega \cdot m$, $h_2 = 60 \, m$, $\rho_3 = 500 \, \Omega \cdot m$;
Q-type model: $\rho_1 = 25 \, \Omega \cdot m$, $h_1 = 60 \, m$, $\rho_2 = 5 \, \Omega \cdot m$, $h_2 = 60 \, m$, $\rho_3 = 1 \, \Omega \cdot m$.

Figure 4.2 shows the results of wave field transformation for the above three models. Analyzing the results, we find that the common feature is that the waveform is broad, the time interval between the positive and negative pulse is big. This results in the reducing resolution on electrical interfaces in the process of pseudo-seismic imaging.

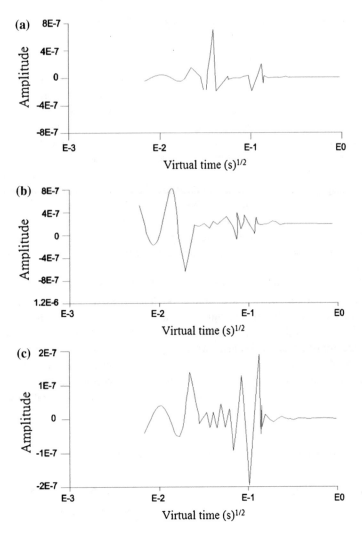

Fig. 4.2 Wave field transformation results for different geoelectric models. **a**, **b**, and **c** are, respectively, for K-, A-, and Q-type model

4.2.2 The Reason of Waveform Dispersion

In this section, we investigate the cause for waveform dispersion. After the TEM field is transformed into the pseudo-seismic wave field, the waveform of the obtained pseudo-seismic wave field will be broadened with increasing time, which will influence the resolution of the imaging. This phenomenon can be attributed to the kernel function $a(t, \tau)$ used in the wave field transform, i.e.,

$$a(t, \tau) = \frac{1}{2\sqrt{\pi t}} \tau e^{-\tau^2/4t}. \tag{4.3}$$

Obviously, the variance of the kernel function with the Gauss distribution is the real time, thus with increasing time, the wave field form becomes more dispersive. Although the waveform dispersion is caused by the mathematical transformation, but it is very similar to the filtering of the earth in a seismic. From the filtering point of view, we need to obtain an inverse filtering coefficient to reduce the filtering influence from the earth to enhance the resolution. This can be done through the deconvolution. It has been theoretically and practically proved that the energy of the dispersion waves can be concentrated, so that the resolution to fine layers can be obtained. To illustrate the increase of distribution width of the kernel function with time, we calculate according to Eq. (2.30) the distributions of the kernel function with virtual time τ for different times t = 80 μs, 325 μs, 800 μs, 2.1 ms, 8.7 ms, 27 ms, 81 ms.

From Fig. 4.3, we can see that when t = 80 μs, the maximum range of τ related to the kernel function is $0 \sim 0.05 \sqrt{s}$; when t = 325 μs, it is $0 \sim 0.1 \sqrt{s}$; when t = 800 μs, it is $0 \sim 0.15 \sqrt{s}$; when t = 2.1 ms, it is $0 \sim 0.25 \sqrt{s}$; when t = 8.7 ms, it is $0 \sim 0.5 \sqrt{s}$; when t = 27 ms, it is $0 \sim 0.85 \sqrt{s}$; when t = 81 ms, it is $0 \sim 1.5 \sqrt{s}$. With increasing time, the distribution range of the kernel function $a(t, \tau)$ obviously becomes broader. This results in the dispersion of the virtual wave field with time after inverse transformation.

Table 4.1 lists the statistics on kernel distribution at different time. It can be seen from the table that the scope of the kernel distribution increases with increasing time. This further indicates that over time the scope of the kernel function distribution increases, the width of virtual wave widens.

Besides, the speed of electromagnetic wave is $c_0 = 1/\sqrt{\mu_0 \varepsilon_0}$ in the free space, while the speed of virtual wave is $v = 1/\sqrt{\mu_0 \sigma}$ in conductive media, the difference can reach 6 orders. This means that due to very slow propagation of virtual wave in conductive media, for the same traveling distance, the pseudo-seismic wave needs more time. The remarkable waveform dispersion is observed.

4.2.3 Waveform Dispersion Compression Based on Deconvolution

The waveform dispersion is caused by the mathematical transformation, which is similar to the earth filtering in a seismic. In the seismic survey, the sharp pulse series indicating the stratigraphic sequence are often blurred by the filtering of the earth, the noise, and multiple waves. Thus, the vertical resolution to the stratigraphic structures is decreased. The deconvolution is to compress the wavelets. By

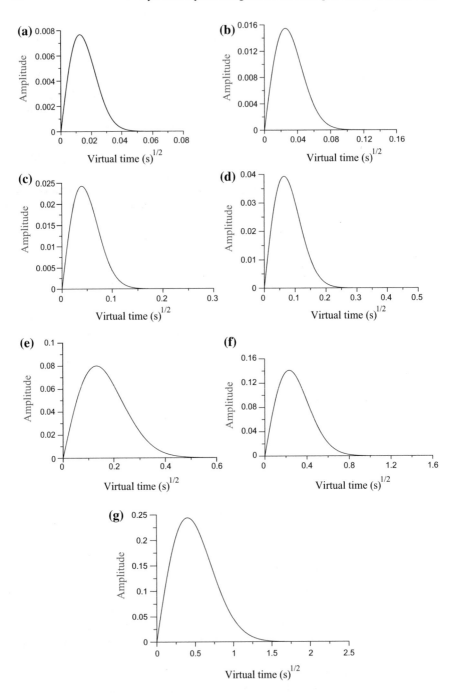

Fig. 4.3 Distribution of the kernel function. **a–g** are for time t = 80 μs, 325 μs, 800 μs, 2.1 ms, 8.7 ms, 27 ms, 81 ms, respectively

Table 4.1 Statistic on the distribution of kernel function

No.	Time (t)	Range of the virtual time τ	Width of the virtual time τ
1	80 μs	$0 \sim 0.05 \sqrt{s}$	$0.05 \sqrt{s}$
2	325 μs	$0 \sim 0.1 \sqrt{s}$	$0.1 \sqrt{s}$
3	800 μs	$0 \sim 0.15 \sqrt{s}$	$0.15 \sqrt{s}$
4	2.1 ms	$0 \sim 0.25 \sqrt{s}$	$0.25 \sqrt{s}$
5	8.7 ms	$0 \sim 0.5 \sqrt{s}$	$0.5 \sqrt{s}$
6	27 ms	$0 \sim 0.85 \sqrt{s}$	$0.85 \sqrt{s}$
7	81 ms	$0 \sim 1.5 \sqrt{s}$	$1.5 \sqrt{s}$

searching for an anti-filtering factor that can eliminate the effect of the earth filtering, we can centralize the energy of the previously dispersed wave field and achieve fine layer distinguishment, so that the resolution can be greatly improved.

Peacock and Treitel (1969) proposed the predictive deconvolution. This technique has been widely used in seismic data processing. Similarly, this technique can also be applied to process the pseudo-seismic wave in the TEM.

Assuming that $y(t)$ is the transformed pseudo-seismic waves, $y(t) = (y(0), y(1), \ldots y(n))$, we hope to create a filter $h(t) = (h(-m_0), h(-m_0 + 1), h(-m_0 + 2), \ldots, h(-m_0 + m))$, where $-m_0$ is the starting time of $h(t)$ and the duration is (m + 1), to run the following calculation

$$\hat{y}_i(t + a) = \sum_{t=1}^{n} h(t) y_i(n - t), \tag{4.4}$$

to obtain the wavelet $\hat{y}(t + a)$ after a delay time a that make the fitting error Q between the calculation values and the pseudo-seismic waves minimum, i.e.,

$$Q = [y(t + a) - \hat{y}(t + a)]^2. \tag{4.5}$$

Substituting Eq. (4.4) into the Eq. (4.5), we obtain

$$Q = \sum_{i=1}^{m} \left[y_i(t + a) - \sum_{i=1}^{m} h(t) y_i(t - t) \right]^2. \tag{4.6}$$

To minimize Q, we need to calculate the partial derivative of Q with respect to the filtering coefficients and make it zero, that is

$$\frac{\partial Q}{\partial h(t)} = 0 \tag{4.7}$$

Insertion of Eq. (4.6) into (4.7), we have

$$
\frac{\partial Q}{\partial h(t)} = \sum_{t=-m_0}^{-m_0+m+n} \frac{\partial Q}{\partial a(l)} \left[\sum_{\tau=m_0}^{-m_0+m} a(\tau)b(t-\tau) - d(t) \right]^2 = 2 \sum_{t=m_0}^{-m_0+m+n} \left[\sum_{\tau=-m_0}^{-m_0+m} a(\tau)b(t-\tau) - d(t) \right] b(t-1)
$$

$$
= 2 \sum_{\tau=-m_0}^{-m_0+m} h(\tau) \sum_{t=-m_0}^{-m_0+m+n} y(t-\tau)y(t-l) - 2 \sum_{t=-m_0}^{-m_0+m+n} \hat{y}(t)y(t-l) = 0,
$$

$$
(l = -m_0, -m_0 + 1, \ldots, -m_0 + m)
$$

$$(4.8)$$

where $\sum_{t=-m_0}^{-m_0+m+n} y(t-\tau)y(t-l) = r_{bb}(l-\tau)$ is the self-correlation of the virtual wavelets, while $\sum_{t=-m_0}^{-m_0+m+n} \hat{y}(t)y(t-l) = r_{db}(l)$ is the cross-correlation between the wavelets and the expected output. Equation (4.8) can be written as

$$
\sum_{\tau=-m_0}^{-m_0+m} h(\tau)r_{bb}(l-\tau) = r_{db}(l) \quad (l = -m_0, -m_0 + 1, \ldots, -m_0 + m). \tag{4.9}
$$

In matrix format, Eq. (4.9) can be expressed as

$$
\begin{bmatrix} r_{bb}(0) & r_{bb}(1) & \cdots & r_{bb}(m) \\ r_{bb}(1) & r_{bb}(0) & \cdots & r_{bb}(m-1) \\ \vdots & \vdots & & \vdots \\ r_{bb}(m) & r_{bb}(m-1) & \cdots & r_{bb}(0) \end{bmatrix} \begin{bmatrix} h(-m_0) \\ h(-m_0+1) \\ \vdots \\ h(-m_0+m) \end{bmatrix} = \begin{bmatrix} h(-m_0) \\ r_{bd}(-m_0+1) \\ \vdots \\ r_{bd}(-m_0+m) \end{bmatrix},
$$

$$(4.10)$$

Solving Eq. (4.10), we can obtain the required deconvolution filter $h(t)$. Then, we can calculate the filtered pseudo-seismic wave $\hat{y}(t)$ by using Eq. (4.4).

Theoretically, the length of the filter can be chosen freely. In seismic, the length is generally chosen to be 80, 120, 160, 200, or 240 ms. However, the length of the filter must be converted into the number of samples. The length should be selected according to the total number of data after interpolation. Only when the length is appropriate, the deconvolution becomes stable. In this study, we choose the length of the deconvolution filter to be 400–600 ms.

4.2.4 Model Calculations

Figure 4.4 shows the deconvolution results. Figure 4.4a displays the waveform before the deconvolution, the waveform of the virtual reflection wave at the interface is obviously wide and the resolution is poor; Fig. 4.4b displays the waveform after deconvolution. Two sets of virtual reflection waves appear, and the waveform is relatively narrow. It is seen that the waveform after the deconvolution

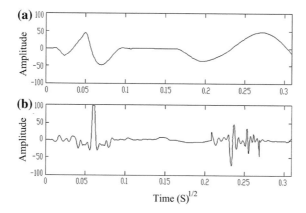

Fig. 4.4 Geological model and comparison of its pseudo-seismic waves before and after deconvolution processing. **a** Waveform before deconvolution. **b** Waveform after deconvolution (H-type of earth model: $\rho_1 = \rho_3 = 100\ \Omega \cdot m$, $\rho_2 = 5\ \Omega \cdot m$, $h_1 = h_2 = 100\ m$.)

is compressed and sharpened so that the boundaries of the underground geoelectrical layers become clearer. This indicates that the deconvolution process on the pseudo-seismic wave can greatly enhance the resolution capability of TEM signals.

To check the effect of deconvolution, we carry out the pseudo-seismic imaging on the virtual wave field after the deconvolution for the model in Fig. 4.4, while Fig. 4.5 shows the imaging results. In theory, the two interfaces of the geoelectrical model are located in the depth of 100 and 200 m, respectively. From Fig. 4.5, we can see that the wave energy produced by the shallower geoelectrical interface lies at approximately 100 m, while the wave energy produced by the deeper interface lies between 190 and 200 m. This indicates that the positions of the geoelectrical interfaces are in good agreement with the real positions 100 and 200 m of the model.

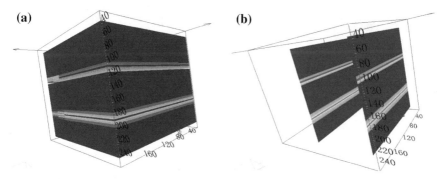

Fig. 4.5 Image for an H-type earth model based on the pseudo-seismic imaging. **a** 3D image; **b** Sliced image

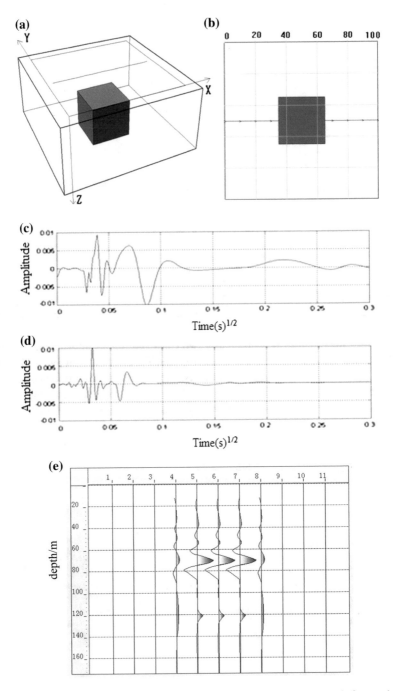

Fig. 4.6 Three-dimensional geological model and its pseudo-seismic wave before and after deconvolution. **a** 3D geological model; **b** Plane view of the calculation area; **c** Waveform before deconvolution; **d** Waveform after deconvolution; **e** Pseudo-seismic imaging section

We design a 3D geoelectrical model to further check the effect of our deconvolution process. The half space has a resistivity of $\rho_1 = 10\,\Omega \cdot m$, the target is
buried 70 m deep with dimensions of 30 m $\times$ 30 m $\times$ 50 m and a resistivity of
$\rho_2 = 300\,\Omega \cdot m$. Figure 4.6 shows the model results. From the figure, we can see
that the waveform is sharpened after the anti-filtering process, the wave energy of
the top layer concentrates at the depth of 70 m, while the wave energy of the lower
boundary concentrates at the depth of 120 m. This is in accordance with the true
model. The wave dispersion effect is more serious for the lower interface of the
target because there exists a weak reflection from high to low resistivity. The
original wave is broadened. However, the waveform after the deconvolution process becomes sharpened and the geological–electrical boundary is more clearly
distinguished.

Reference

Peacock KL, Treitel S (1969) Predictive deconvolution-theory and practices. Geophysics
34(2):155

Chapter 5
Surface Continuation and Imaging of TEM Based on Pseudo Wave Equations

In the previous chapter, we have solved the problem of TEM wave field transformation and proved that there indeed exists superposition of multi-aperture TEM field through experimental research. These create conditions for TEM migration and imaging. However, there exist similarity and difference between TEM migration and imaging and the migration and imaging of elastic wave in seismic. If we transform the TEM diffusion field into the wave field and use the method for wave field to interpret TEM data, namely the method using reverse extrapolation from ground to underground to do the migration and imaging for the seismic wave. This method is called TEM migration and imaging that uses Kirchhoff integral to realize the continuation calculation of EM wave. This is the focus and frontier research in TEM method. We will discuss the details of this method in three sections.

5.1 Establishment of Kirchhoff Diffraction Integral

As shown in Fig. 5.1, we choose an area W with the boundary Q, $\mathbf{n}$ is the outward normal of Q. At the moment t, the potential function observed at point $M(x_1, y_1, z_1)$ is $\varphi(x_1, y_1, z_1, t_1)$, then at the previous moment $t_1 = t - \frac{r}{v}$, the displacement potential at the closed surface Q is $\varphi(x_1, y_1, z_1, t - \frac{r}{v}) = [\varphi]$. We called $[\varphi]$ the retarded potential, r is the distance from point M to any point at Q, namely $r = \sqrt{(x - x_1)^2 + (y - y_1)^2 + (z - z_1)^2}$.

To calculate the total displacement potential on Q, we use the Green's theorem

$$\int_W (u\nabla^2 v - v\nabla^2 u)\, \mathrm{d}w = \oint_Q \left(u\frac{\partial v}{\partial n} - v\frac{\partial u}{\partial n}\right) \mathrm{d}Q, \qquad (5.1)$$

where u and v are arbitrary functions, we select $u = \varphi(x, y, z, t - \frac{r}{v}) = [\varphi]$.

© Science Press and Springer Nature Singapore Pte Ltd. 2017
X. Li et al., *Migration Imaging of the Transient Electromagnetic Method*,
DOI 10.1007/978-981-10-2708-6_5

Fig. 5.1 The model and
boundaries for the
establishment of Kirchhoff
diffraction integral

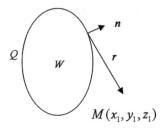

We need the retarded potential $[\varphi]$ at $t_1 = t - \frac{r}{v}$ to obtain the solution of φ at the moment of t. We choose $v = \frac{1}{r}$, since $\varphi = \frac{1}{v}C_1\left(t - \frac{r}{v}\right)$, while $\frac{1}{r}$ is the basic solution of φ, so spherical diffusion of wavelet depends on $\frac{1}{r}$. Inserting $v = \frac{1}{r}$ into the left-hand side of Eq. (5.1) and considering that $\nabla^2\left(\frac{1}{r}\right) = 0$, we have Eq. (5.1) reformulated to

$$-\int_W \frac{1}{r}\nabla^2[\varphi]dw = \oint_Q \left[[\varphi]\frac{\partial}{\partial n}\left(\frac{1}{r}\right) - \frac{1}{r}\frac{\partial}{\partial n}\cdot\frac{\partial[\varphi]}{\partial r}\right]dQ, \tag{5.2}$$

With $\frac{\partial}{\partial n} = \frac{\partial}{\partial r}\cdot\frac{\partial r}{\partial n}$, the above equation can be changed into

$$\int_W \frac{1}{r}\nabla^2[\varphi]dw = \oint_Q \left[\frac{[\varphi]}{r^2}\frac{\partial r}{\partial n} + \frac{1}{r}\cdot\frac{\partial r}{\partial n}\frac{\partial[\varphi]}{\partial r}\right]dQ. \tag{5.3}$$

In Eq. (5.3), $\nabla^2[\varphi] = [\nabla^2\varphi]$, while

$$[\nabla^2\varphi] = [\nabla\cdot\nabla\varphi] = [\nabla\cdot E], \tag{5.4}$$

where $E = \nabla\varphi$, $\varphi = \frac{1}{r}C_1\left(t - \frac{r}{v}\right)$, thus we have $[\nabla^2\varphi] = [\theta]$.

In above equations, $[\theta]$ is relative change of volume of excitation source at the moment of $t_1 = t - \frac{r}{v}$. According to the definition of the excitation source intensity, the volume integration at the left side of Eq. (5.3) is the intensity of excitation source. The surface integration of Green's theorem is the solution at surface Q due to the source excitation. The solution at Q acts as a new excitation sources for the secondary wavelets, while the surface integration at Q is the summation of the secondary wavelets produced by the new excitation sources on Q. The meaning of Eq. (5.3) is that in model region W enclosed by Q the energy produced by primary source at $t_1 = t - \frac{r}{v}$ is equal to summation of quadratic wavelet on Q. Thus, Eq. (5.3) is the key for obtaining the solution of displacement potential excited by excitation source. In the following, we build the integral formulation using $[\varphi]$ and its derivative:

1. Using deviation of composite functions

$$\frac{\partial u}{\partial x} = \frac{\partial}{\partial x} \varphi\left(x, y, z, t - \frac{r}{v}\right) = \left(\frac{\partial}{\partial x} - \frac{1}{v}\frac{\partial r}{\partial x} \cdot \frac{\partial}{\partial t_1}\right) \varphi\left(x, y, z, t - \frac{r}{v}\right). \quad (5.5)$$

When $t = t_1$, then

$$\frac{\partial u}{\partial x} = \left[\frac{\partial \varphi}{\partial x}\right] - \frac{1}{v}\frac{\partial r}{\partial x}\left[\frac{\partial \varphi}{\partial t_1}\right], \quad (5.6)$$

$$\frac{\partial u}{\partial n} = \left[\frac{\partial \varphi}{\partial n}\right] - \frac{1}{v}\frac{\partial r}{\partial n}\left[\frac{\partial \varphi}{\partial t_1}\right], \quad (\partial t_1 = \partial t). \quad (5.7)$$

2. Solving $\nabla^2 u$

By expanding the Laplace operator, we obtain

$$\nabla^2 u = \frac{\partial^2 u}{\partial x^2} + \frac{\partial^2 u}{\partial y^2} + \frac{\partial^2 u}{\partial z^2}. \quad (5.8)$$

From (5.6), we obtain $\frac{\partial^2 u}{\partial x^2}$ as

$$\frac{\partial^2 u}{\partial x^2} = \frac{\partial}{\partial x}\left[\frac{\partial \varphi}{\partial x}\right] - \frac{1}{v}\frac{\partial r}{\partial x}\frac{\partial}{\partial x}\left[\frac{\partial \varphi}{\partial t}\right] - \frac{1}{v}\frac{\partial^2 r}{\partial x^2}\frac{\partial}{\partial x}\left[\frac{\partial \varphi}{\partial t}\right]. \quad (5.6)$$

Since

$$\frac{\partial}{\partial x}\left[\frac{\partial \varphi}{\partial x}\right] = \left(\frac{\partial}{\partial x} - \frac{1}{v}\frac{\partial r}{\partial x} \cdot \frac{\partial}{\partial t}\right) \cdot \left[\frac{\partial \varphi}{\partial x}\right] = \left[\frac{\partial^2 \varphi}{\partial x^2}\right] - \frac{1}{v}\frac{\partial r}{\partial x}\frac{\partial}{\partial x}\left[\frac{\partial^2 \varphi}{\partial t \partial x}\right], \quad (5.7)$$

$$\frac{\partial}{\partial x}\left[\frac{\partial \varphi}{\partial t}\right] = \left(\frac{\partial}{\partial x} - \frac{1}{v}\frac{\partial r}{\partial x} \cdot \frac{\partial}{\partial t}\right)\left[\frac{\partial \varphi}{\partial t}\right] = \left[\frac{\partial^2 \varphi}{\partial x \partial t}\right] - \frac{1}{v}\frac{\partial r}{\partial x} \cdot \left[\frac{\partial^2 \varphi}{\partial t^2}\right]. \quad (5.8)$$

Substitution of (5.7) and (5.8) into (5.6) yields

$$\frac{\partial^2 u}{\partial x^2} = \left[\frac{\partial^2 \varphi}{\partial x^2}\right] - \frac{2}{v}\frac{\partial r}{\partial x}\left[\frac{\partial^2 \varphi}{\partial x \partial t}\right] + \frac{1}{v^2}\left(\frac{\partial r}{\partial x}\right)^2\left[\frac{\partial^2 \varphi}{\partial t^2}\right] - \frac{1}{v}\frac{\partial^2 r}{\partial t^2}\left[\frac{\partial \varphi}{\partial t}\right]. \quad (5.9)$$

Similarly, we have

$$\frac{\partial^2 u}{\partial y^2} = \left[\frac{\partial^2 \varphi}{\partial y^2}\right] - \frac{2}{v}\frac{\partial r}{\partial y}\left[\frac{\partial^2 \varphi}{\partial y \partial t}\right] + \frac{1}{v^2}\left(\frac{\partial r}{\partial y}\right)^2\left[\frac{\partial^2 \varphi}{\partial t^2}\right] - \frac{1}{v}\frac{\partial^2 r}{\partial y^2}\left[\frac{\partial \varphi}{\partial t}\right], \quad (5.10)$$

$$\frac{\partial^2 u}{\partial z^2} = \left[\frac{\partial^2 \varphi}{\partial z^2}\right] - \frac{2}{v}\frac{\partial r}{\partial z}\left[\frac{\partial^2 \varphi}{\partial z \partial t}\right] + \frac{1}{v^2}\left(\frac{\partial r}{\partial z}\right)^2\left[\frac{\partial^2 \varphi}{\partial t^2}\right] - \frac{1}{v}\frac{\partial^2 r}{\partial z^2}\left[\frac{\partial \varphi}{\partial t}\right]. \tag{5.11}$$

Thus,

$$\nabla^2 u = [\nabla^2 \varphi] - \frac{2}{v}\sum_{i-1}^{3}\frac{\partial r}{\partial W_i}\left[\frac{\partial^2 \varphi}{\partial W_i \partial t}\right] + \frac{1}{v^2}\left[\frac{\partial^2 \varphi}{\partial t^2}\right]\sum_{i-1}^{3}\left[\frac{\partial r}{\partial W_i}\right]^2$$
$$- \frac{1}{v}\left[\frac{\partial \varphi}{\partial t}\right]\sum_{i-1}^{3}\left[\frac{\partial^2 r}{\partial W_i^2}\right], \tag{5.12}$$

where $W_1 = x$, $W_2 = y$, $W_3 = z$. Since $r^2 = (x - x_1)^2 + (y - y_1)^2 + (z - z_1)^2$, we have

$$\frac{\partial r}{\partial x} = \frac{x - x_1}{r}, \quad \frac{\partial r}{\partial y} = \frac{y - y_1}{r}, \quad \frac{\partial r}{\partial x} = \frac{z - z_1}{r}, \tag{5.13}$$

$$\sum_{i-1}^{3}\left(\frac{\partial r}{\partial W_i}\right)^2 = \left(\frac{x - x_1}{r}\right)^2 + \left(\frac{y - y_1}{r}\right)^2 + \left(\frac{z - z_1}{r}\right)^2 = 1, \tag{5.14}$$

$$\sum_{i-1}^{3}\left(\frac{\partial r^2}{\partial W_1^2}\right) = \frac{\partial}{\partial x}\left(\frac{\partial r}{\partial x}\right) + \frac{\partial}{\partial y}\left(\frac{\partial r}{\partial y}\right) + \frac{\partial}{\partial z}\left(\frac{\partial r}{\partial z}\right) = \frac{2}{r}. \tag{5.15}$$

Besides, we have the wave equation

$$\nabla \varphi = \frac{1}{v^2}\frac{\partial^2 \varphi}{\partial t^2} + F(x, y, z, t). \tag{5.16}$$

Substituting all equations above into (5.12), we obtain

$$\nabla^2 u = \frac{2}{v^2}\left[\frac{\partial^2 \varphi}{\partial t^2}\right] - \frac{2}{vr}\left[\frac{\partial \varphi}{\partial t}\right] - \frac{2}{v}\sum_{i-1}^{3}\frac{\partial r}{\partial W_i}\left[\frac{\partial^2 \varphi}{\partial W_i \partial t}\right] + F. \tag{5.17}$$

Thus, the integrand $\frac{1}{r}\nabla^2 u$ at the left-hand side of Green's theorem can be written as

$$\frac{1}{r}\nabla^2 u = \frac{2}{v^2 r}\left[\frac{\partial^2 \varphi}{\partial t^2}\right] - \frac{2}{vr^2}\left[\frac{\partial \varphi}{\partial t}\right] - \frac{2}{v}\sum_{i-1}^{3}\frac{W_i - W_{1i}}{r^2}\left[\frac{\partial^2 \varphi}{\partial W_i \partial t}\right] + \frac{F}{r}. \tag{5.18}$$

To make the above equation more meaningful and understandable, we run the following formulation:

$$\frac{\partial}{\partial x}\left\{\frac{x-x_i}{r^2}\left[\frac{\partial\varphi}{\partial t}\right]\right\} = \frac{\partial}{\partial x}\left(\frac{x-x_1}{r^2}\right)\cdot\left[\frac{\partial\varphi}{\partial t}\right] + \frac{x-x_1}{r^2}\frac{\partial}{\partial x}\left[\frac{\partial\varphi}{\partial t}\right]$$

$$= \frac{1}{r^2}\left[\frac{\partial\varphi}{\partial t}\right] - \frac{2(x-x_1)^2}{r^4}\left[\frac{\partial\varphi}{\partial t}\right] + \frac{x-x_1}{r^2}\left[\frac{\partial^2\varphi}{\partial x\partial t}\right] - \frac{(x-x_1)^2}{vr^3}\left[\frac{\partial^2\varphi}{\partial x\partial t}\right]. \tag{5.19}$$

Similarly, we can obtain $\frac{\partial}{\partial y}\left\{\frac{y-y_1}{r^2}\left[\frac{\partial\varphi}{\partial t}\right]\right\}$ and $\frac{\partial}{\partial z}\left\{\frac{z-z_1}{r^2}\left[\frac{\partial\varphi}{\partial t}\right]\right\}$.

Summation of them yields

$$\sum_{i-1}^{3}\frac{\partial}{\partial W_i}\left\{\frac{W_i-W_{1i}}{r^2}\left[\frac{\partial\varphi}{\partial t}\right]\right\} = \frac{1}{r^2}\left[\frac{\partial\varphi}{\partial t}\right] - \frac{1}{vr}\left[\frac{\partial^2\varphi}{\partial t^2}\right] + \sum_{i-1}^{3}\frac{W_i-W_{1i}}{r^2}\left[\frac{\partial^2\varphi}{\partial W_i\partial t}\right],$$

$$\tag{5.20}$$

where $W_1 = x$, $W_2 = y$, $W_3 = z$. The above equation, multiplied by $-\frac{2}{v}$ and then added with $\frac{F}{r}$, is equal to Eq. (5.18). Then

$$\frac{1}{r}\nabla^2 u = -\frac{2}{v}\left(\frac{\partial f_x}{\partial x} + \frac{\partial f_y}{\partial y} + \frac{\partial f_z}{\partial z}\right) + \frac{F}{r} = -\frac{2}{v}\nabla\cdot\vec{f} + \frac{F}{r}. \tag{5.21}$$

Assuming that

$$\frac{x-x_1}{r^2}\left[\frac{\partial\varphi}{\partial t}\right] = f_x, \quad \frac{y-y_1}{r^2}\left[\frac{\partial\varphi}{\partial t}\right] = f_y, \quad \frac{z-z_1}{r^2}\left[\frac{\partial\varphi}{\partial t}\right] = f_z, \tag{5.22}$$

Then (5.21) can be written as

$$\frac{1}{r}\nabla^2 u = -\frac{2}{v}\left(\frac{\partial f_x}{\partial x} + \frac{\partial f_y}{\partial y} + \frac{\partial f_z}{\partial z}\right) + \frac{F}{r} = -\frac{2}{v}\nabla\cdot f + \frac{F}{r}. \tag{5.23}$$

According to divergence theorem

$$\int_W \nabla\cdot f\,dw = \oint_Q f_n dQ, \tag{5.24}$$

we can obtain

$$\int_W \frac{1}{r}\nabla^2 u\,dw = \int_W -\frac{2}{v}\nabla\cdot f\,dw + \int_W \frac{F}{r}\,dw = -\frac{2}{v}\oint_Q f_n dQ + \frac{\mu_0\delta(t-o^*)}{r_0}. \tag{5.25}$$

Since $f_n = \frac{1}{r}\frac{\partial r}{\partial n}\left[\frac{\partial \varphi}{\partial t}\right]$, we rewrite the above equation as

$$\int_W \frac{1}{r}\nabla^2 u\,dw - \frac{F}{r_0} = -\frac{2}{r}\oint_Q \frac{1}{r}\frac{\partial r}{\partial n}\left[\frac{\partial \varphi}{\partial t}\right]dQ, \qquad (5.26)$$

where r_0 is the distance between the excitation source to reference points. Substituting Eqs. (5.22) and (5.7) into Green's theorem and considering that $u = [\varphi]$, for point M located outside Q, the left-hand side of Green's theorem is

$$\int_W (u\nabla^2 v - v\nabla^2 u)dw = -\int_W \frac{1}{v}\nabla^2 u\,dw = \frac{2}{v}\oint_Q \frac{1}{r}\frac{\partial r}{\partial n}\left[\frac{\partial \varphi}{\partial t}\right]dQ + \frac{F}{r_0}, \qquad (5.27)$$

while the right-hand side of Green's theorem is

$$\oint_Q \left(u\frac{\partial v}{\partial n} - v\frac{\partial u}{\partial n}\right)dQ = \oint_Q \left[[\varphi]\frac{\partial r}{\partial n}\frac{\partial}{\partial r}\left(\frac{1}{r}\right) - \frac{1}{r}\frac{\partial}{\partial n}[\varphi]\right]dQ$$

$$= \oint_Q \left\{[\varphi]\left(-\frac{1}{r^2}\right)\frac{\partial r}{\partial n} - \frac{1}{r}\left[\left[\frac{\partial \varphi}{\partial n}\right] - \frac{1}{v}\frac{\partial r}{\partial n}\left[\frac{\partial \varphi}{\partial t}\right]\right]\right\}dQ \qquad (5.28)$$

Then, from the Green's theorem, we have

$$\frac{2}{v}\oint_Q \frac{1}{v}\frac{\partial r}{\partial n}\left[\frac{\partial \varphi}{\partial t}\right]dQ + \frac{F}{r_0} = \oint_Q -\frac{1}{r^2}[\varphi]\frac{\partial r}{\partial n}dQ - \oint_Q \frac{1}{r}\left[\frac{\partial \varphi}{\partial n}\right]dQ + \frac{1}{v}\oint_Q \frac{1}{r}\frac{\partial r}{\partial n}\left[\frac{\partial \varphi}{\partial t}\right]dQ, \qquad (5.29)$$

The Kirchhoff integral can be expressed as

$$\oint_Q \left\{-\frac{[\varphi]}{r^2}\frac{\partial r}{\partial n} - \frac{1}{r}\left[\frac{\partial \varphi}{\partial n}\right] - \frac{1}{r}\frac{\partial r}{\partial n}\left[\frac{\partial \varphi}{\partial t}\right]\right\}dQ = \frac{F}{r_0}, \qquad (5.29)$$

where $F = \mu_0\delta(t - o^*)$. At point M, $\varphi = \frac{F}{r_0}$

3. Point M located within disturbing area W

When $r \to 0$, $v = \frac{1}{r} \to \infty$, the Green's theorem fails. We need to remove the singularity point M. We design a sphere σ around M with a radius ε and apply Green's theorem in the area between surface Q and σ, then from Eq. (5.29) we obtain

$$\oint_{Q+\sigma} \left\{ [\varphi] \frac{\partial}{\partial n} \left(\frac{1}{r} \right) - \frac{1}{r} \left[\frac{\partial \varphi}{\partial n} \right] - \frac{1}{vr} \frac{\partial r}{\partial n} \left[\frac{\partial \varphi}{\partial t} \right] \right\} dQ = \frac{F}{r_0}. \tag{5.30}$$

Taking into account that

$$\iint_{\sigma} \left\{ [\varphi] \frac{\partial}{\partial n} \left(\frac{1}{\varepsilon} \right) - \frac{1}{\varepsilon} \left[\frac{\partial \varphi}{\partial n} \right] - \frac{1}{v\varepsilon} \frac{\partial \varepsilon}{\partial n} \left[\frac{\partial \varphi}{\partial t} \right] \right\} d\sigma, \tag{5.31}$$

as ε is very small, we have $\frac{\partial}{\partial n} = -\frac{\partial}{\partial \varepsilon}, d\sigma = \varepsilon^2 d\Omega, d\Omega$ is the solid angle, the above equation becomes

$$\iint_{\sigma} \left\{ -[\varphi] \frac{\partial}{\partial \varepsilon} \left(\frac{1}{\varepsilon} \right) + \frac{1}{\varepsilon} \left[\frac{\partial \varphi}{\partial \varepsilon} \right] + \frac{1}{v\varepsilon} \left(\frac{\partial \varepsilon}{\partial \varepsilon} \right) \left[\frac{\partial \varphi}{\partial t} \right] \right\} d\sigma$$

$$\underset{d\sigma = \varepsilon^2 d\Omega}{=} \iint_{\Omega} \left\{ [\varphi] + \varepsilon \left[\frac{\partial \varphi}{\partial \varepsilon} \right] + \frac{\varepsilon}{v} \left[\frac{\partial \varphi}{\partial t} \right] \right\} d\Omega \qquad , \tag{5.32}$$

$$\underset{\varepsilon \to 0}{=} \iint_{\Omega} [\varphi] d\Omega = 4\pi[\varphi]$$

Substitution into (5.30) yields

$$\iint_{Q} \left\{ [\varphi] \frac{\partial}{\partial n} \left(\frac{1}{r} \right) - \frac{1}{r} \left[\frac{\partial \varphi}{\partial n} \right] - \frac{1}{vr} \frac{\partial r}{\partial n} \left[\frac{\partial \varphi}{\partial t} \right] \right\} dQ + 4\pi\varphi = \frac{F}{r_0}. \tag{5.33}$$

Thus, the displacement potential at point M in area W is

$$\varphi(x_M, y_M, z_M, t) = -\frac{1}{4\pi} \iint_{Q} \left\{ [\varphi] \frac{\partial}{\partial n} \left(\frac{1}{r} \right) - \frac{1}{r} \left[\frac{\partial \varphi}{\partial n} \right] - \frac{1}{vr} \frac{\partial r}{\partial n} \left[\frac{\partial \varphi}{\partial t} \right] \right\} dQ + \frac{F}{r_0}, \tag{5.34}$$

where $F = \mu_0 \delta(t - o^*)$. Only when $t = 0$, $F(0) = \mu_0$; otherwise $F(t) = 0$. In general, $t \neq 0$, we do not need to consider the term $F(0) = \mu_0$.

5.2   Migration by Kirchhoff Integration (Surface Continuation)

As we know, we can use the wave equation to describe the wave field propagating in the underground. The equation for longitudinal wave is

$$\nabla^2 \mu - \frac{1}{v^2} \frac{\partial^2 u}{\partial t^2} = F, \tag{5.35}$$

Fig. 5.2 The model and
boundaries for the migration
by Kirchhoff integration

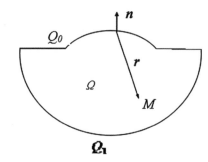

and the solution in Kirchhoff integration is

$$u(x,y,z,t) = -\frac{1}{4\pi} \oiint_Q \left\{ [u]\frac{\partial}{\partial n}\left(\frac{1}{r}\right) - \frac{1}{r}\left[\frac{\partial u}{\partial n}\right] - \frac{1}{vr}\frac{\partial r}{\partial n}\left[\frac{\partial u}{\partial t}\right] \right\} dQ + \frac{F}{r_0}, \quad (5.36)$$

where $F = \mu_0\delta(t - 0^x)$, $Q = Q_0 + Q_1$ is a close surface. As shown in Fig. 5.2, Q_0 is the earth surface, Q_1 is infinite semi-spherical surface.

When $r \to \infty$, $u \to 0$, $\frac{\partial u}{\partial t} \to 0$, the surface integration on Q has no contribution to wave field function at point $M(X, Y, Z)$, so that

$$\iint_Q \left\{ [u]\frac{\partial}{\partial n}\left(\frac{1}{r}\right) - \frac{1}{r}\left[\frac{\partial u}{\partial n}\right] - \frac{1}{vr}\frac{\partial r}{\partial n}\left[\frac{\partial u}{\partial t}\right] \right\} dQ = 0. \quad (5.37)$$

Thus, Eq. (5.36) can be written as

$$u(x,y,z,t) = -\frac{1}{4\pi} \iint_Q \left\{ [u]\frac{\partial}{\partial n}\left(\frac{1}{r}\right) - \frac{1}{r}\left[\frac{\partial u}{\partial n} - \frac{1}{vr}\frac{\partial r}{\partial n}\left[\frac{\partial u}{\partial t}\right]\right] \right\} dQ + \frac{F}{r_0} \quad (5.38)$$

The Migration of pseudo-seismic wave field is an inverse process of obtaining seismic record. What we have got is the pseudo-seismic records at the earth surface for electromagnetic responses. We need to determine the spatial locations of virtual secondary sources at the reflection surfaces. For $u(x, y, z, t)$, when t turn into $-t$, $W(x, y, z, t) = u(x, y, z, -t)$ will satisfy the same wave equation. If it is a forward time problem for $W(x, y, z, t)$, it is then an backward time problem for $u(x, y, z, t)$. We can take the points at the reflection surfaces as source points that can induce up-going waves. In this way, we can take the surface receiving points as secondary excitation source and shift backward the time signal to its original state to find the wave function and to determine the reflection surfaces.

Suppose a zero-offset up-going wave $G(x, y, z_0, t)$, it is the value of wave field $g(x, y, z, t)$ at the earth surface $z = z_0$ that is induced by the secondary sources at the underground reflection surfaces. From (5.38), we have

$$g(x,y,z,t) = -\frac{1}{4\pi} \iint\limits_{Q_0} \left\{ \frac{\partial}{\partial n}\left(\frac{1}{r}\right) - \frac{1}{r}\frac{\partial}{\partial n} - \frac{1}{vr}\frac{\partial r}{\partial n}\frac{\partial}{\partial t} \right\} G\left(\xi,\eta,\zeta_0,t+\frac{r}{v}\right) dQ + \frac{F}{r_0},$$

$$(5.39)$$

where $G\left(\xi,\eta,\zeta_0,t+\frac{r}{v}\right)$ takes the time $t+\frac{r}{v}$, this is because we consider the inverse process of wave field. This means that we calculate the wave field in the underground from its surface value $G(\xi,\eta,\zeta_0,t)$ to determine the reflection surfaces. This is the downward continuation of the wave field.

5.3   Boundary Element Method for Wave Field Continuation

5.3.1   Discretization of Kirchhoff Integration

We divide the boundary Q_0 with n nodes into elements and calculate the integral over the boundary in Eq. (5.39) by summing the integrals over each element. For point p within W, we have

$$g(x,y,z,t) = -\frac{1}{4\pi} \iint\limits_{Q_0} \left\{ \frac{\partial}{\partial n}\left(\frac{1}{r}\right) - \frac{1}{r}\frac{\partial}{\partial n} - \frac{1}{vr}\frac{\partial r}{\partial n}\frac{\partial}{\partial t} \right\} G\left(\xi,\eta,\zeta_0,t+\frac{r}{v}\right) dQ + \frac{F}{r_0}.$$

$$(5.40)$$

5.3.2   Analysis on Elements

Refer to Fig. 5.3, we assume the triangle element has vertex of j, k, m, their coordinates are (x_j, y_j, z_j), (x_k, y_k, z_k), (x_m, y_m, z_m). We first carry out the coordinate transformation and denote the coordinate using shape function ξ_j, ξ_k, ξ_m, i.e.,

Fig. 5.3 Triangle element

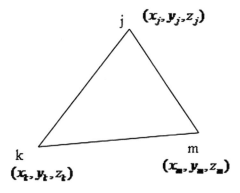

$$
\begin{matrix}
x = x_j\xi_j + x_k\xi_k + x_m\xi_m \\
y = y_j\xi_j + y_k\xi_k + y_m\xi_m \\
z = z_j\xi_j + z_k\xi_k + z_m\xi_m
\end{matrix}
\Rightarrow \cdots \cdot
\begin{bmatrix}
x_j, x_k, x_m \\
y_j, y_k, y_m \\
z_j, z_k, z_m
\end{bmatrix}
\cdot
\begin{bmatrix}
\xi_j \\
\xi_k \\
\xi_m
\end{bmatrix}
=
\begin{bmatrix}
x \\
y \\
z
\end{bmatrix},
\qquad (5.41)
$$

where $\xi_j + \xi_k + \xi_m = 1$ have the values ranging from 0 to 1. $\xi_j + \xi_k + \xi_m = 1$ and at point j, $\xi_j = 1$, $\xi_k = \xi_m = 0$; at point k, $\xi_j = 1$, $\xi_k = \xi_m = 0$; while at point m, $\xi_j = 1$, $\xi_k = \xi_m = 0$, thus ξ_j, ξ_k, ξ_m are linear functions of x, y, z.

Since element Q_ε is generally very small, we can assume that the wave field G varies linearly within the element. Then, G is written as

$$
G = \xi_j G_j + \xi_k G_k + \xi_m G_m = [\xi_j, \xi_k, \xi_m]
\begin{bmatrix}
G_j \\
G_k \\
G_m
\end{bmatrix},
\qquad (5.42)
$$

where G_j, G_k, G_m is value of G at the surface point Q_0.

Considering the integral at an element

$$
\iint_{\Gamma_e} \frac{\partial}{\partial n}\left(\frac{1}{r}\right) G\left(\xi, \eta, \zeta_0, t + \frac{r}{v}\right) d\Gamma
$$
$$
= \iint_{\Gamma_e} -\frac{1}{r^2}\frac{\partial}{\partial n} G\left(\xi, \eta, \zeta_0, t + \frac{r}{v}\right) d\Gamma
\qquad (5.43)
$$
$$
= -\iint_{\Gamma_e} \frac{\cos(r\hat{\cdot}n)}{r^2} G\left(\xi, \eta, \zeta_0, t + \frac{r}{v}\right) d\Gamma
$$

where $\frac{\partial r}{\partial n} = \cos(r\hat{\cdot}n)$. We denote the wave field in the element $G\left(\xi, \eta, \zeta_0, t + \frac{r}{v}\right)$ by G_e, then

$$
-\iint_{\Gamma_e} \frac{\cos(r\hat{\cdot}n)}{r^2} G\left(\xi, \eta, \zeta_0, t + \frac{r}{v}\right) d\Gamma
$$
$$
= -\sum_{e=j,k,m} \iint_{\Gamma_e} \frac{\cos(r\hat{\cdot}n)}{r^2} d\Gamma \cdot G_e
\qquad = -\sum_{e=j,k,m} f_{i.e.} G_e,
\qquad (5.44)
$$

where

$$
f_{ie} = \iint_{\Gamma_e} \xi_e \frac{\cos(r\hat{\cdot}n)}{r^2} dT \quad (e = j, k, m).
$$

The above equation can be calculated by Gauss integration, i.e.,

$$f_{ie} = \iint_{\Gamma_e} \xi_e \frac{\cos(\hat{r \cdot n})}{r^2} d\Gamma = \sum_{q=1}^{4} \xi_e(q) \frac{\cos(\hat{r \cdot n})}{r^2} W_g \cdot \Delta, \tag{5.45}$$

where W_g is the weighted coefficients, Δ is the triangle area.

Similar to above formulation, we can handle the following element integrations:

$$\iint_{\Gamma_e} \frac{1}{r} \frac{\partial G}{\partial n} d\Gamma = \sum_{e=j,k,m} d_{ie} \frac{\partial G_e}{\partial n}, \tag{5.46}$$

with

$$d_{ie} = \iint_{\Gamma_e} \xi_e \frac{1}{r} d\Gamma = \sum_{q=1}^{4} \xi_e(q) \frac{1}{r} W_q \cdot \Delta, \tag{5.47}$$

and

$$\iint_{\Gamma_e} \frac{1}{vr} \frac{\partial r}{\partial n} \frac{\partial G}{\partial t} d\Gamma = \sum_{i=j,k,m} c_{ie} \frac{\partial G_e}{\partial t}, \tag{5.48}$$

with

$$c_{ie} = \iint_{\Gamma_e} \xi_e \frac{\cos(\hat{r \cdot n})}{vr} = \sum_{q=1}^{4} \xi_e(q) \frac{\cos(\hat{r \cdot n})}{vr} W_q \cdot \Delta. \tag{5.49}$$

5.3.3 Total Matrix

Substituting Eqs. (5.44), (5.46), (5.48), into (5.39), we obtain the wave field at point i

$$g_i = -\frac{1}{4\pi} \left\{ F_i G - D_i \frac{\partial G}{\partial n} - C_i \frac{\partial G}{\partial t} \right\}, \tag{5.50}$$

For the total of n nodes, we obtain an equations system

$$\mathbf{G} = -\frac{1}{4\pi} \left\{ \mathbf{F} \cdot \mathbf{G} - \mathbf{D} \frac{\partial \mathbf{G}}{\partial \mathbf{n}} - \mathbf{C} \frac{\partial \mathbf{G}}{\partial \mathbf{t}} \right\}, \tag{5.51}$$

where $F = \lfloor F_{ij} \rfloor, D = \lfloor D_{ij} \rfloor, C = \lfloor C_{ij} \rfloor$. F_{ij} is the sum of f_{ij} at adjacent elements of node j, D_{ij} is the sum of d_{ij} at adjacent elements of node j, while C_{ij} is the sum of c_{ij} at adjacent elements of node j. If the wave field at the earth surface

$G = \{g_1, g_2, \ldots g_n\}^T$, the normal derivative $\frac{\partial G}{\partial n} = \left\{\frac{\partial g_1}{\partial n}, \frac{\partial g_2}{\partial n}, \ldots\ldots, \frac{\partial g_n}{\partial n}\right\}^T$, and the velocity $\frac{\partial G}{\partial t} = \left\{\frac{\partial g_1}{\partial t}, \frac{\partial g_2}{\partial t}, \ldots\ldots, \frac{\partial g_n}{\partial t}\right\}^T$ are known, we can calculate the wave field at any underground points.

5.3.4 Integration Over Elements

(1) Manipulation of $\cos(\hat{r} \cdot \boldsymbol{n})$

As discussed above, $\cos(\hat{r} \cdot \boldsymbol{n}) = \pm D/r$, D is distance from node i to the plane of an element, r is the distance from node i to an arbitrary point on the plane. In the following, we first determine D.

Supposing that the surface element is located in a plane of

$$ax + by + cz + 1 = 0, \tag{5.52}$$

then substitution of the coordinates of three nodes into it yields

$$\begin{cases} ax_j + by_j + cz_j + 1 = 0 \\ ax_k + by_k + cz_k + 1 = 0 \\ ax_m + by_m + cz_m + 1 = 0 \end{cases}, \tag{5.53}$$

or in matrix format

$$\begin{bmatrix} x_j & y_j & z_j \\ x_k & y_k & z_k \\ x_m & y_m & z_m \end{bmatrix} \cdot \begin{bmatrix} a \\ b \\ c \end{bmatrix} = -\begin{bmatrix} 1 \\ 1 \\ 1 \end{bmatrix} \tag{5.54}$$

for the solution of a, b, c. r_D is distance from i to arbitrary point on the plane with $r_D^2 = (x - x_i)^2 + (y - y_i)^2 + (z - z_i)^2$.

From Eq. (5.52), we have

$$z = -\frac{ax + by + 1}{c}. \tag{5.54}$$

Substituting it into r_D, taking the derivative with respect to x and y, respectively, and making it equal to zero, we obtain

$$\left. \begin{aligned} \frac{\partial r_D^2}{\partial x} &= 2(x - x_i) + 2\left(\frac{ax + by + 1}{c} + z_i\right)\frac{a}{c} = 0 \\ \frac{\partial r_D^2}{\partial y} &= 2(y - y_i) + \left(\frac{ax + by + 1}{c} + z_i\right)\frac{b}{c} = 0 \end{aligned} \right\}. \tag{5.55}$$

After solving x and y, denoted by x_D and y_D, substituting them into Eq. (5.52), we can get z_D, then

$$\mathbf{D} = \mathbf{i}(x_D - x_i) + \mathbf{j}(y_D - y_i) + \mathbf{k}(z_D - z_i), \tag{5.56}$$

$$|\mathbf{D}| = \sqrt{(x_D - x_i)^2 + (y_D - y_i)^2 + (z_D - z_i)^2}. \tag{5.57}$$

Thus,

$$\cos(\hat{\mathbf{r} \cdot \mathbf{n}}) = \frac{|\mathbf{D}|}{|\mathbf{r}|}, \tag{5.58}$$

where $|\mathbf{r}| = \sqrt{(x - x_i)^2 + (y - y_i)^2 + (z - z_i)^2}$, point (x, y, z) is on the plane. Next, we derive x_D, y_D, z_D. From Eq. (5.55), we have

$$(x - x_i) + \left(\frac{ax + by + 1}{c} + z_i\right)\frac{a}{c} = 0, \tag{5.59}$$

$$(y - y_i) + \left(\frac{ax + by + 1}{c} + z_i\right)\frac{b}{c} = 0. \tag{5.60}$$

Multiplication of (5.59) and (5.60), respectively by a/c and b/c and subtraction of them yields

$$\frac{a}{c}(x - x_i) - \frac{c}{b}(y - y_i) = 0, \tag{5.61}$$

so that

$$x = \frac{a(y - y_i) + bx_i}{b}. \tag{5.62}$$

Substituting (5.62) into (5.58), we have

$$\left(\frac{a(y - y_i) + bx_i}{b} - x_i\right) + \left(\frac{a\frac{[a(y-y_i)+by+1]}{b} + by + 1}{c} + z_i\right)\frac{a}{c} = 0, \tag{5.63}$$

so that

$$y = \frac{(c^2 + a^2)y_i - bax_i - bcz_i - b}{a^2 + b^2 + c^2}. \tag{5.64}$$

Substituting (5.64) into (5.62), we have

$$
\begin{aligned}
x &= \frac{1}{b}\left[a\left(\frac{(c^2 + a^2)y_i - bax_i - bcz_i - b}{a^2 + b^2 + c^2} - y_i \right) + bx_i \right] \\
&= \frac{(b^2 + c^2)x_i - aby_i - acz_i - a}{a^2 + b^2 + c^2},
\end{aligned}
\tag{5.65}
$$

Substituting (5.64) and (5.65) into (5.54), we have

$$
z = -\frac{ax + by + 1}{c} = \frac{(a^2 + b^2)z_i - acx_i - bcy_i - c}{a^2 + b^2 + c^2}.
\tag{5.66}
$$

We denote x, y, z in Eqs. (5.64)–(5.66) by z_D, z_D and z_D, i.e.,

$$
\left.
\begin{aligned}
x_D &= \frac{(b^2 + c^2)x_i - aby_i - acz_i - a}{a^2 + b^2 + c^2} \\
y_D &= \frac{(c^2 + a^2)y_i - abx_i - baz_i - b}{a^2 + b^2 + c^2} \\
z_D &= \frac{(a^2 + b^2)z_i - acx_i - bcy_i - c}{a^2 + b^2 + c^2}
\end{aligned}
\right\}.
\tag{5.67}
$$

(2) Calculation of integral over elements

We take the integral f_{ij} as example. From the above discussion, we have

$$
f_{ij} = \iint_{T_e} \xi_j \frac{\cos(\boldsymbol{r}\hat{\cdot}\boldsymbol{n})}{r^2} dr = \sum_{q=1}^{4} \xi_j(q) \frac{\cos(\boldsymbol{r}\hat{\cdot}\boldsymbol{n})}{r^2} w_g \Delta.
\tag{5.68}
$$

Using four-point Gauss integration, refer to Table 5.1, we have

$$
\begin{aligned}
f_{ij} &= \frac{1}{3}\frac{\cos(\boldsymbol{r}_1\hat{\cdot}\boldsymbol{n})}{r_1^2} \cdot (-9/16)\Delta + \frac{3}{5}\frac{\cos(\boldsymbol{r}_2\hat{\cdot}\boldsymbol{n})}{r_2^2}(25/48)\Delta \\
&+ \frac{1}{5}\frac{\cos(\boldsymbol{r}_3\hat{\cdot}\boldsymbol{n})}{r_3^2}(25/48)\Delta + \frac{1}{5}\frac{\cos(\boldsymbol{r}_4\hat{\cdot}\boldsymbol{n})}{r_4^2}(25/48)\Delta.
\end{aligned}
\tag{5.69}
$$

Table 5.1 Coefficients of four-point Gauss integration

q	$\xi_j^{(q)}$	$\xi_k^{(q)}$	$\xi_m^{(q)}$	w_g
1	1/3	1/3	1/3	−9/16
2	3/5	1/5	1/5	25/48
3	1/5	3/5	1/5	25/48
4	1/5	1/5	3/5	25/48

In the following, we first calculate the triangle area Δ. From Fig. 5.3, we have

$$\left.\begin{array}{l} \boldsymbol{r}_{kj} = \boldsymbol{i}(x_k - x_j) + \boldsymbol{j}(y_k - y_j) + \boldsymbol{k}(z_k - z_j) \\ \boldsymbol{r}_{mk} = \boldsymbol{i}(x_m - x_k) + \boldsymbol{j}(y_m - y_k) + \boldsymbol{k}(z_m - z_k) \end{array}\right\}, \tag{5.70}$$

while

$$\begin{aligned} \boldsymbol{r}_{kj} \cdot \boldsymbol{r}_{mk} &= |\boldsymbol{r}_{kj}||\boldsymbol{r}_{mk}| \cdot \cos(180 - \theta) \\ &= -|\boldsymbol{r}_{kj}||\boldsymbol{r}_{km}| \cos\theta, \end{aligned} \tag{5.71}$$

where the direction of $\boldsymbol{r}_{kj}$ is from node j to k, while the direction $\boldsymbol{r}_{mk}$ is from node k to m, θ is the angel between $\boldsymbol{r}_{kj}$ and $\boldsymbol{r}_{mk}$. Thus, we have

$$\cos\theta = -\frac{(x_k - x_j)(x_m - x_k) + (y_k - y_j)(y_m - y_k) + (z_k - z_j)(z_m - z_k)}{\sqrt{(x_k - x_j)^2 + (y_k - y_j)^2 + (z_k - z_j)^2}\sqrt{(x_m - x_k)^2 + (y_m - y_k)^2 + (z_m - z_k)^2}}, \tag{5.72}$$

The height and area of triangular element are respectively

$$h = \sqrt{(x_k - x_j)^2 + (y_k - y_j)^2 + (z_k - z_j)^2}\,\sin\theta, \tag{5.73}$$

and

$$\begin{aligned} \Delta &= \frac{1}{2}h|\boldsymbol{r}_{mk}| = \frac{1}{2}|\boldsymbol{r}_{kj}||\boldsymbol{r}_{mk}|\sin\theta = \frac{1}{2}|\boldsymbol{r}_{kj}||\boldsymbol{r}_{mk}|\sqrt{1 - \left(\frac{\boldsymbol{r}_{kj} \cdot \boldsymbol{r}_{mk}}{|\boldsymbol{r}_{kj}||\boldsymbol{r}_{mk}|}\right)^2} \\ &= \frac{1}{2}\sqrt{(|\boldsymbol{r}_{kj}||\boldsymbol{r}_{mk}|)^2 - [(x_k - x_j)(x_m - x_k) + (y_k - y_j)(y_m - y_k) + (z_k - z_j)(z_m - z_k)]^2} \end{aligned} \tag{5.74}$$

After obtaining the area of the triangular element, we now calculate $|\boldsymbol{r}|$. For this purpose, we rewrite $r_1\,r_2\,r_3\,r_4$ from Eq. (5.69) as

$$r_q = \sqrt{(x_q - x_i)^2 + (y_q - y_i)^2 + (z_q - z_i)^2}.. \tag{5.75}$$

In above equation, q is the node number of Gauss integration, r_d is the distance from point (x, y, z) in domain Ω to an arbitrary point (x_q, y_q, z_q) on the element. We know that the coordinate on the triangular element can be expressed as

$$\left.\begin{array}{l} x = x_j \xi_k^{(q)} + x_k \xi_k^{(q)} + x_m \xi_m^{(q)} \\[4pt] y = y_j \xi_j^{(q)} + y_k \xi_k^{(q)} + y_m \xi_m^{(q)} \\[4pt] z = z_j \xi_j^{(q)} + z_k \xi_k^{(q)} + z_m \xi_m^{(q)} \end{array}\right\}. \tag{5.76}$$

For four-point Gauss integration ($q = 1, 2, 3, 4$), when $q = 1$, we have

$$r_1 = \sqrt{(x_1 - x_i)^2 + (y_1 - y_i)^2 + (z_1 - z_i)^2}, \tag{5.77}$$

with

$$\begin{cases} x_1 = x_j \xi_j^{(q)} + x_k \xi_k^{(q)} + x_m \xi_m^{(q)} = \frac{1}{3} x_j + \frac{1}{3} x_k + \frac{1}{3} x_m \\[4pt] y_1 = y_j \xi_j^{(q)} + y_k \xi_k^{(q)} + y_m \xi_m^{(q)} = \frac{1}{3} y_j + \frac{1}{3} y_k + \frac{1}{3} y_m \\[4pt] z_1 = z_j \xi_j^{(q)} + z_k \xi_k^{(q)} + z_m \xi_m^{(q)} = \frac{1}{3} z_j + \frac{1}{3} z_k + \frac{1}{3} z_m \end{cases}. \tag{5.78}$$

When $q = 2, 3, 4$, we have

$$r_2 = \sqrt{(x_2 - x_i)^2 + (y_2 - y_i)^2 + (z_2 - z_i)^2}, \tag{5.79}$$

with

$$\begin{cases} x_2 = \frac{3}{5} x_j + \frac{1}{5} x_k + \frac{1}{5} x_m \\[4pt] y_2 = \frac{3}{5} y_j + \frac{1}{5} y_k + \frac{1}{5} y_m \\[4pt] z_2 = \frac{3}{5} z_j + \frac{1}{5} z_k + \frac{1}{5} z_m \end{cases}, \tag{5.80}$$

$$r_3 = \sqrt{(x_3 - x_i)^2 + (y_3 - y_i)^2 + (z_3 - z_i)^2}, \tag{5.81}$$

With

$$\begin{cases} x_3 = \frac{1}{5} x_j + \frac{3}{5} x_k + \frac{1}{5} x_m \\[4pt] y_3 = \frac{1}{5} y_j + \frac{3}{5} y_k + \frac{1}{5} y_m \\[4pt] z_3 = \frac{1}{5} z_j + \frac{3}{5} z_k + \frac{1}{5} z_m \end{cases}. \tag{5.82}$$

$$r_4 = \sqrt{(x_4 - x_i)^2 + (y_4 - y_i)^2 + (z_4 - z_i)^2}, \tag{5.83}$$

with

$$
\begin{cases}
x_4 = \frac{1}{5}x_j + \frac{1}{5}x_k + \frac{3}{5}x_m \\
y_4 = \frac{1}{5}y_j + \frac{1}{5}y_k + \frac{3}{5}y_m \\
z_4 = \frac{1}{5}z_j + \frac{1}{5}z_k + \frac{3}{5}z_m
\end{cases}. \tag{5.84}
$$

Chapter 6
Velocity Analysis of TEM Pseudo Wave Field

With the application of pseudo wave field migration method, the velocity modelling technology that directly impacts the quality of migration imaging attracts more and more attention. In seismic prospecting, velocity modelling technology can be divided into methods based on ray theory and wave equations. The wave equation method can be divided into depth focusing analysis and residual curvature analysis, while the ray method is based on ray tracing theory. According to the channel gathers, we can divide it into common central point (CMP), common imaging point (CIP), common reflection point (CRP), common focusing point (CFP), and common reflection surface gathers (CRS). These method can be implemented in TEM pseudo wave field depth migration. However, we need to note that TEM pseudo wave field is obtained through mathematical transform, which is not objective one. Thus, TEM pseudo wave field is has its own characteristics. In this chapter, starting from the characteristic of TEM field, we put forward continuous velocity analysis for pseudo wave field based on the theory of equivalent conductive plate.

6.1 Velocity Modelling Based on Equivalent Conductive Plate Method

Equivalent conductive plate method is an approximate method that intuitively divides layers based on the features of vertical apparent conductance curves. This method can be vividly understood as: with increasing or decreasing time t, the equivalent conductive plane floats up and down with the velocity of $\frac{1}{\mu_0\sigma}$. Thus, we can replace a uniform half-space under the loop source by a conductive plate floating up and down with the time, and calculate the EM field at the earth surface. The floating speed of conductive plate is only connected with the earth's conductivity, similar to the case of pseudo wave field. Thus, we can use the equivalent

© Science Press and Springer Nature Singapore Pte Ltd. 2017
X. Li et al., *Migration Imaging of the Transient Electromagnetic Method*,
DOI 10.1007/978-981-10-2708-6_6

conductive plate method to find out the distribution of the vertical conductivity, namely to calculate the vertical conductivities for different depths. Then we can calculate velocity from the equation of TEM pseudo wave field and realize the continuous velocity analysis.

6.1.1 Basic Theory of Equivalent Conductive Plate Method

The theory of equivalent conductive plate method was first developed by Russian scientist in 1970s. For completeness of the algorithm development, we summarize the method in this section. We assume a circular loop at the earth surface powered by a step current and we cut the power at the time $t = 0$, i.e.

$$I(t) = \begin{cases} I & t<0 \\ 0 & t\geq 0 \end{cases}. \tag{6.1}$$

Then, at the time $t \geq 0$, the eddy current will be produced in the underground, the EM field induced by the eddy current can be observed at any point on the ground. According to the EM theory, we can use a conductive plate to replace the homogeneous medium in the underground and calculate the induced EM field at any point. Refer to Fig. 6.1, we assume that A is a completely conductive plate with conductance of σ, the transmitting loop has a radius of a, the transmitting current is I.

When the power is cut-off, there will be eddy current produced in conductive plate A. To calculate the EM field induced by the eddy current, we use a virtual source located at the symmetric location h to replace the eddy current in the plate. In this way, we can obtain the EM field induced by the virtual source.

We introduce a vector potential A, then the electric field at any point in the conductive plate is

$$E = -\frac{\partial A}{\partial t}, \tag{6.2}$$

Fig. 6.1 An equivalent conductive plate model

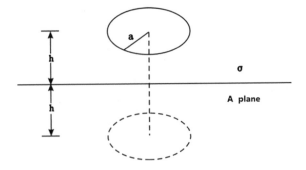

while the surface current density is

$$\frac{j}{\sigma_{\text{surface}}} = \frac{\partial A}{\partial t}. \tag{6.3}$$

Based on the distribution of eddy current in the plane, the vector potential A has only a component A_φ that is a function of r. The eddy current is also a function of r. Assuming that the total current between any point r and the outside boundary of the plate is $I(r)$, the surface current density is

$$j = \frac{\mathrm{d}I(r)}{\mathrm{d}r}. \tag{6.4}$$

According to Ampere's law,

$$\mu_0 I(r) = \oint_L \boldsymbol{B} \cdot \mathrm{d}\boldsymbol{l}, \tag{6.5}$$

where the path of integration L include the segments running from r to the outside boundary above the plate, bypassing the boundary and returning back to point r. Then, from Eq. (6.5), we have

$$\mu_0 j = \mu_0 \frac{\mathrm{d}I(r)}{\mathrm{d}r} = \frac{\mathrm{d}}{\mathrm{d}r} \oint_L \boldsymbol{B} \cdot \mathrm{d}\boldsymbol{l} = 2 \frac{\mathrm{d}}{\mathrm{d}r} \int_0^\infty Br \, \mathrm{d}r = -2Br, \tag{6.6}$$

and

$$j = -\frac{2Br}{\mu_0}. \tag{6.7}$$

Substituting (6.7) into (6.3), we have

$$\frac{2Br}{\sigma_{\text{surface}} \mu_0} = \frac{\partial A_\varphi}{\partial t}. \tag{6.8}$$

From $\boldsymbol{B} = \nabla \times \boldsymbol{A}$, we have $B_r = -\frac{\partial A_\varphi}{\partial z}$. Substitution into (6.8) yields

$$\frac{2}{\mu_0 \sigma_{\text{surface}}} \frac{\partial A_\varphi}{\partial z} = -\frac{\partial A_\varphi}{\partial t} \quad (\sigma = \sigma_{\text{surface}}). \tag{6.9}$$

The solution for A_φ at any point outside the plate can be expressed as

$$A_\varphi = f\left(Z + \frac{2t}{\mu_0 \sigma}\right). \tag{6.10}$$

Based on the image method, the potential A_φ induced by imaging source can be expressed as

$$A_\varphi = f\left(z + \frac{2t}{\mu_0 \sigma}\right)_{z=2h} = f\left(2h + \frac{2t}{\mu_0 \sigma}\right). \tag{6.11}$$

This shows that when $t > 0$, the virtual source moves down at a speed of $2/\mu_0 \sigma$. This can be interpreted as follow: with time t increasing or decreasing, the equivalent conductive plate floats up and down at a speed of $1/\mu_0 \sigma$. When t is increasing, the equivalent conductive plate floats up, when t is decreasing, the equivalent conductive plate floats down. In this way, we can use a current-bearing floating plate to replace the conductive earth, so that we can calculate the EM field at the earth surface.

6.1.2   Approximate Calculation of TEM Field at the Surface of a Horizontally Layered Earth

In TEM method, apparent vertical conductivity refers to as the vertical conductivity of a uniform conductive plate at certain depth that can replace the conductivity of the earth in the effective scope of EM field. The equivalent conductive plate and its vertical conductivity change with time. In the following, we derive the expression for the anomalous field at the loop center.

In the polar coordinate system, we assume that the current is I, the virtual source coincides with x, y-plane, and the center of the virtual source is origin. The vertical component of magnetic field induced by a linear current element Ids is

$$dH_z = \frac{1}{4\pi} I \frac{ds}{R^2} \sin\theta, \tag{6.12}$$

the magnetic field induced by the whole loop is

$$H_z = \frac{1}{4\pi} \oint_L \frac{ds}{R^2} \sin\theta = \frac{Ia^2}{2} \frac{1}{(a^2 + z^2)^{3/2}}. \tag{6.13}$$

Considering that equivalent conductive plate is floating, we substitute $z = 2h + 2t/\mu_0 \sigma$ into (6.13) and obtain

$$H_z = \frac{Ia^2}{2} \frac{1}{\left[a^2 + 4\left(h + \frac{t}{\mu_0 \sigma}\right)\right]^{3/2}}, \qquad (6.14)$$

thus

$$\frac{\partial B_z}{\partial t} = -\frac{6Ia^2}{s} \frac{h + \frac{t}{\mu_0 s}}{\left[a^2 + 4\left(h + \frac{t}{\mu_0 s}\right)^2\right]^{5/2}}, \qquad (6.15)$$

where $s = \sigma$ is the conductivity of equivalent conductive plate.

Equation (6.14) and (6.15) is the approximate calculation of TEM field at center point of the loop at the surface of a uniform half-space with equivalent conductive plate method.

Assuming

$$m = h + \frac{t}{\mu_0 s}, \qquad (6.16)$$

and

$$\bar{m} = \frac{h}{a} + \frac{t}{\mu_0 sa}, \qquad (6.17)$$

thus

$$H_z = \frac{I}{2a} \frac{a^3}{[a^2 + 4m^2]^{3/2}} = \frac{I}{2a} \frac{1}{[1 + 4\bar{m}^2]^{3/2}}, \qquad (6.18)$$

$$\frac{\partial B_z}{\partial t} = -\frac{6I}{sa^2} \frac{a^4 m}{[a^2 + 4m^2]^{5/2}} = -\frac{6I}{sa^2} \frac{\bar{m}}{[1 + 4\bar{m}^2]^{5/2}}. \qquad (6.19)$$

When we further assume that $4m^2 \gg 1$ or $a^2 \ll 4m^2$, then the approximate expression for late-time TEM field can be expressed as

$$H_z = \frac{Ia^2}{16m^3}, \qquad (6.20)$$

$$\frac{\partial B_z}{\partial t} = -\frac{3Ia^2}{16sm^4}. \qquad (6.21)$$

For horizontally layered strata, we can also obtain the approximate solution of TEM field by using the equivalent conductive plate method, only the floating-down speed of the equivalent conductive plate is no longer constant, but is connected with the conductivity at each layer.

For a horizontally layered medium, we introduce the depth of equivalent conductive plate h_{eq}, equal to average thickness of the layer units that form the section. Then, we have

$$h_{eq} = \left[\frac{\int_0^H \sigma(z)z^g dz}{\int_0^H \sigma(z)dz} \right]^{\frac{1}{g}} = \left[\frac{\int_0^H \sigma(z)z^g dz}{S} \right]^{\frac{1}{g}}, \tag{6.22}$$

where $\sigma(z)$ is the conductivity, H is the depth of investigation, S is total vertical conductivity of the medium above the depth H, g is a parameter used to determine the weights for the upper and lower layers when calculating h_{eq}, with $g > 0$, $0 \leq h_{eq} \leq H$ and h_{eq} is equal to central depth of equivalent conductive plate.

According to Eq. (6.19), the approximate equation for calculating $\frac{\partial B_z(t)}{\partial t}$ with equivalent conductive plate is

$$\frac{\partial B_z(t)}{\partial t} = -\frac{6I}{sa^2} \frac{\bar{m}}{(1+4\bar{m}^2)^{5/2}}. \tag{6.23}$$

This equation can be expressed as

$$\frac{\partial B_z(t)}{\partial t} = \frac{k}{s} F(\bar{m}), \tag{6.24}$$

where

$$k = -\frac{6I}{a^2}, \quad \text{and} \quad F(\bar{m}) = \frac{\bar{m}}{(1+4\bar{m}^2)^{5/2}}. \tag{6.25}$$

Then, we can introduce the apparent vertical conductivity from Eq. (6.24)

$$S_\tau(t) = \frac{KF(\bar{m})}{\frac{\partial B_z(t)}{\partial t}}. \tag{6.26}$$

This equation can be used to process the survey data $\frac{\partial B_z(t)}{\partial t}$.

6.1.3   Optimized Extraction of Parameter $\bar{m}$

According to Eq. (6.26), $F(\bar{m})$ is essential for calculating $S_\tau(t)$. $F(\bar{m})$ is related to the depth h of the equivalent conductive plate or the investigation depth that cannot be directly obtained from the field data. However, we can get $F(\bar{m})$ indirectly by differentiating $\frac{\partial B_z(t)}{\partial t}$. For the survey data, we calculate the derivative with respect to the time, i.e.

$$\frac{\partial^2 B_z}{\partial t^2} = \frac{\partial}{\partial \bar{m}}\left(\frac{\partial B_z}{\partial t}\right) \cdot \frac{\partial(\bar{m})}{\partial t} = \frac{K}{S} \cdot \frac{\partial F(\bar{m})}{\partial \bar{m}} \cdot \frac{\partial t}{\partial(\bar{m})} = \frac{K}{S} \cdot \frac{\partial F(\bar{m})}{\partial m} \cdot \frac{1}{\mu_0 S a}. \tag{6.27}$$

Introducing an auxiliary function $\varphi(\bar{m})$

$$\varphi(\bar{m}) = \frac{\left|\frac{\partial^2 B_z}{\partial t^2}\right|}{\left[\frac{\partial B_z(t)}{\partial t}\right]^2} \cdot \mu_0 a K, \tag{6.28}$$

we have

$$\varphi(\bar{m}) = \frac{\frac{K}{S}\left|\frac{\partial F(\bar{m})}{\partial \bar{m}}\right| \frac{1}{\mu_0 a K}}{\left[\frac{K}{S} F(\bar{m})\right]^2} \mu_0 a K = \left|\frac{\partial F(\bar{m})}{\partial \bar{m}}\right| / [F(\bar{m})]^2. \tag{6.29}$$

Since

$$F(\bar{m}) = \frac{\bar{m}}{(1 + 4\bar{m}^2)^{5/2}}, \tag{6.30}$$

then we have

$$\varphi(\bar{m}) = (1 + 4\bar{m}^2)^{3/2} \left|\frac{1}{\bar{m}^2} - 16\right|. \tag{6.31}$$

Summarizing the above discussions, we can give the steps for obtaining the vertical conductivity:

(1) Get $\frac{\partial B_z(t)}{\partial t}$ and its time derivative;
(2) Substitute $\frac{\partial B_z(t)}{\partial t}$ and $\frac{\partial^2 B_z(t)}{\partial t^2}$ into Eq. (6.28) to get the auxiliary function;
(3) Calculate $\varphi(\bar{m})$ and corresponding $\bar{m}$ through spline interpolation;
(4) Substitute $\bar{m}$ into Eq. (6.30) and (6.26) to get $S_\tau(t)$, and substitute $S_\tau(t)$ into the following equation to get $H_\tau(t)$,

$$H_\tau(t) = (m - t/\mu_0 S_\tau)^{1/3} \left(\frac{9}{8}m - \frac{t}{\mu_0 S_\tau}\right)^{2/3}. \tag{6.32}$$

From these procedures, we can see that $\bar{m}$ is essential for calculating the apparent vertical conductivity S_τ and apparent depth H_τ. $\bar{m}$ can be obtained from Eq. (6.31). $\phi(\bar{m})$ in Eq. (6.31) is only related to $\bar{m}$. We can consider $\phi(\bar{m})$ obtained from Eq. (6.31) as theoretical $\phi(\bar{m})$, written as $\varphi_L(\bar{m})$. $\frac{\partial B_z(t)}{\partial t}$ in step (1) is a set of survey data, the time t is predetermined, while $\frac{\partial^2 B_z(t)}{\partial t^2}$ can be calculated, so we can take the auxiliary function obtained thorough Eq. (6.28) as the observed $\phi(\bar{m})$, written as $\varphi_S(\bar{m})$.

In the above discussion, $\phi(\bar{m})$ in Eq. (6.28) and (6.31) are equal, implying that $\varphi_S(\bar{m})$ is equal to $\varphi_L(\bar{m})$. This create the conditions for us to use the optimization algorithm to solve for $\bar{m}$. In other words, $\bar{m}$ directly influences the final results of apparent vertical conductivity S_τ and apparent depth H_τ. Thus, it is necessary to obtain the optimal value of $\bar{m}$. For this purpose, we introduce the objective function,

$$\Phi(\bar{m}) = \left\| \frac{\left| \frac{\partial^2 B_z}{\partial t^2} \right|}{\left[\frac{\partial B_z(t)}{\partial t} \right]^2} \cdot \mu_0 a K - \left(1 + 4\bar{m}^2 \right)^{3/2} \left| \frac{1}{\bar{m}^2} - 16 \right| \right\|^2 . \tag{6.33}$$

When $\Phi(\bar{m})$ is minimized, the corresponding $\bar{m}$ is the optimal value.

Most optimization algorithm need an initial value and make judgement whether the initial value is the optimal solution. If it is the case, then the optimization is stopped. Otherwise, one needs to adjust this value and repeat the process until it reach an optimal value. The initial value has a big influence on the final result, but how to determine this value is difficult. Algorithms without needs of initial value can be used, such as the artificial intelligence technology or genetic algorithm.

6.2 Velocity Model for Single Observation Point

Figure 6.2 shows the scheme for velocity estimation of pseudo wave field based on equivalent conductive plate method. For observation with zero offset, we start from the top layer. After giving initial parameters $\bar{m}^{(0)}$, we can obtain $s_\tau^{(0)}$ and $h_\tau^{(0)}$.

Adding a small disturbance to $\bar{m}^{(0)}$, we then use Newton iterative method to calculate

$$\bar{m}^{(1)} = \bar{m}^{(0)} + \delta\bar{m}, \tag{6.34}$$

After getting a new model, we use the minimal value of $\Phi(\bar{m})$ to determine the first estimated value $\bar{m}^{(1)}$, then we get the first estimated value $s_\tau^{(1)}$ and $h_\tau^{(1)}$. Repeating these steps, we can get a set of apparent vertical conductivities changing with apparent depth. After getting vertical conductivity, we can obtain conductivity σ via the following Equation:

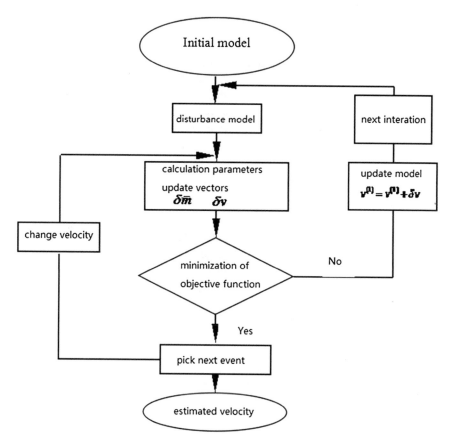

Fig. 6.2 Iteration process for velocity modelling

$$S_\tau(h) = \sum_{i=1}^{N} \sigma_i h_i, \tag{6.35}$$

$$\sigma_i = \frac{S_\tau(h_i) - S_\tau(h_{i-1})}{h_i}, \tag{6.36}$$

and the velocity for pseudo wave field is

$$v = 1/\sqrt{\mu_0 \sigma}. \tag{6.37}$$

6.3 Continuous Velocity Analysis

The limited survey data results in limited amount of velocity v of pseudo wave field obtained from these data. This can be problematic when doing continuation and imaging of pseudo wave field. To solve this problem, we use 3D spatial interpolation method to increase data quantity under the condition that the data quality is not influenced. We try the procedure in Fig. 6.3 to choose optimal interpolation method.

6.3.1 Weighted Interpolation Based on Global Distance

We first use the weighted interpolation method based on global distance to interpolate original pseudo velocity. The weighted interpolation method is based on global distance (???) to interpolate the scattered data. The weights are chosen to be

Fig. 6.3 Selection of spatial interpolation methods

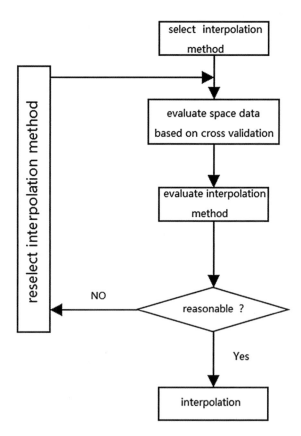

inversely proportional to the distance. The interpolated velocity is affected by the number of data and interpolation points.

We assume that there are a set of discrete points on a plane with known coordinate (x_i, y_i) and physical parameter P_i $(i = 1, 2, \ldots n)$. For global distance weighted interpolation, the physical parameter of grid point $p(x, y)$ is obtained from average value of the points in the local neighborhood, weighted by the distance. Different points have different effects on $P(z)$. This difference is called the weight functions $W_i(x, y)$. $W_i(x, y)$ depends mainly on distance and the direction.

$$P(z) = \frac{\sum_{i=1}^{n} \frac{z_i}{[d_i(x,y)]^n}}{\sum_{i=1}^{n} \frac{1}{[d_i(x,y)]^n}}, \tag{6.38}$$

where

$$d_i(x, y) = \sqrt{(x - x_i)^2 + (y - y_i)^2}. \tag{6.39}$$

$d_i(x, y)$ is the distance between $p(x, y)$ and discrete point (x_i, y_i). $P(z)$ is the value at the to be interpolated.

Weighting function $W_i(x, y)$ is inversely proportional to the uth power of the distance between known points and forecast point. With increasing distance, the weighting decreases rapidly. The decreasing speed is determined by the power u. For a large number u, the points close to the forecast point will be given a large weight. For a small number u, the weights will distribute more uniformly.

Global distance weighted interpolation method creates a rough data surface that is affected by singular values of the data surface (especially the extreme values at the surface edges). At the same time, the density of grids also affects the interpolation results. Figure 6.4a shows the interpolation result of X, Z direction with 30 grids, while Fig. 6.4b shows the interpolation result of X, Z direction with 100 grids. Obviously, the result of Fig. 6.4b is better than that in Fig. 6.4a.

As the first step, we use this method to expand data quantity, while at the same time guarantee the interpolated pseudo-velocity values consistent with the original velocity.

6.3.2 Localized Linear Interpolation

Using localized linear interpolation method to process the data obtained through global distance weighted interpolation, we can obtain the pseudo-velocity at points we need.

First we need to grid the data. To match the extension imaging, we use the triangular mesh similar to the method of the curved surface extension. Refer to Fig. 6.5, the ordinary numbers are node numbers of the grids, the numbers with circle are unit numbers.

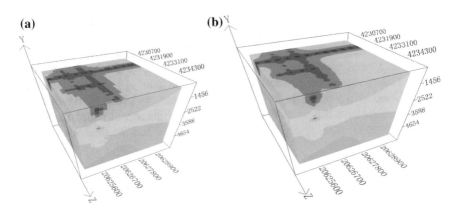

Fig. 6.4 Global distance weighted interpolations with different grids. **a** X, Z direction, 30 grids; **b** X, Z direction, 100 grids

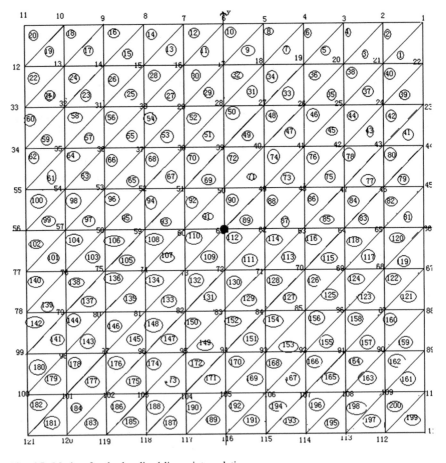

Fig. 6.5 Meshes for the localized linear interpolation

The original file obtained by meshing contains the information of coordinates, depth, and electrical parameters of every point.

Then, we will define the neighborhood of each interpolation point. There exist certain relevance between the two sampling points close to each other that decreases with the distance. When the distance is long enough, the sampling point and forecast point are not related. Thus, we need in practice to define the shape of the neighborhood and number of known sampling points.

For a good forecasting scheme, the observation points should come from a number of directions. Refer to Fig. 6.6, a circular neighborhood is divided into four sections, and in each section we have 3–5 forecasting points.

The sampling position have serious effects on the spatial interpolation. When the data are related and uniformly distributed, they can better demonstrate the distribution character of research objects, and thus can help better in the data analysis and selection of interpolation method. If data points swarm or don't distribute uniformly, the interpolation results will be inaccurate. Thus, the sampling points should distribute as uniformly as possible.

Refer to Fig. 6.7, for each point to be interpolated, all known points above it will affect its velocity value. Thus, every known point at the earth surface will form a neighborhood together with the interpolation point. To calculate the size of the neighborhood, we define the line running through the known point and interpolation point and take it as the diagonal line of a cube to build the neighborhood. Every known data point in the neighborhood is the observation point. We choose the nearest point from the interpolation points and take its value as the temporary value of the interpolation point. We repeat this procedure by shifting to the next point at the earth surface, until all known points at the earth surface have been used. Finally, we take the average of all temporary values as the final value of the interpolation point and save the final result into a file. This finish a localized linear interpolation.

To follow the Green's theorem, we must guarantee that the element integration for the continuation follows the anti-clockwise direction, Thus, in the interpolation process, the cyclic sequence of surface known points must be consistent with node number, as shown in Fig. 6.8.

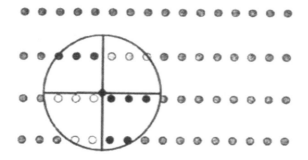

Fig. 6.6 Definition of forecasting observation points in neighbor

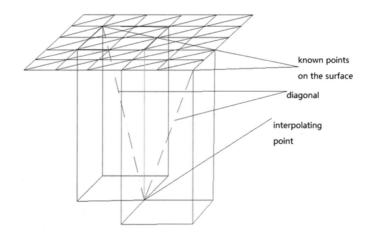

Fig. 6.7 Sketch for localized linear interpolation

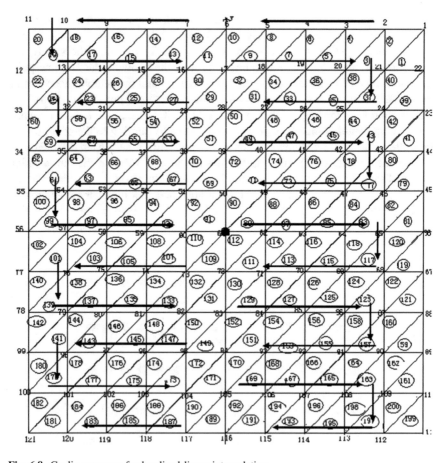

Fig. 6.8 Cyclic sequence for localized linear interpolation

Fig. 6.9 Near point linear interpolation sketch

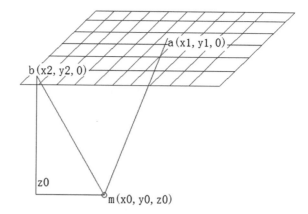

In the above process, a key step is to determine the diagonal of the neighborhood. Refer to Fig. 6.9, assuming that the coordinates of point a, b, m are known, then the equations for am or bm can be written as

$$\frac{x - x_0}{m} = \frac{y - y_0}{n} = \frac{z - z_0}{p}, \tag{6.40}$$

where $m = x_2 - x_0$, $n = y_2 - y_0$, $p = z_2 - z_0$. According to the equation above, parameter equations for the lines are

$$\begin{cases} x = x_0 + mt \\ y = y_0 + nt \\ z = z_0 + pt. \end{cases} \tag{6.41}$$

For the data points required for wave field continuation, we can predetermine the distance for the downward continuation, and thus z is known. Substituting z into above Eq. (6.34), we can get t, and then x, y values. After this, we can look for the nearest point to the target point in the scope of cube. The virtual velocity of this point is taken as the virtual velocity of target point. The flow chart of localized linear interpolation is shown in Fig. 6.10.

The accuracy of this 3D spatial interpolation is determined by the quantity of original data. This interpolation method is linear interpolation, so its accuracy is lower than spline interpolation or Kriging interpolation. However, it is much faster than the other methods. This makes it very applicable in the practical use.

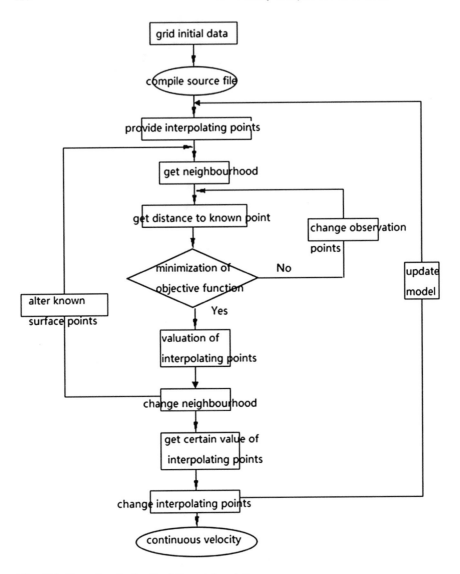

Fig. 6.10 Flow chart for localized linear interpolation

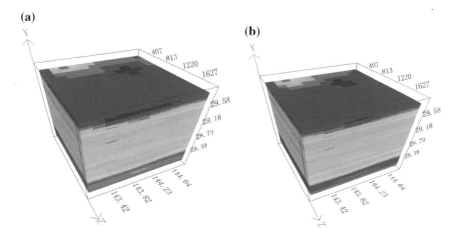

Fig. 6.11 The localized linear interpolation results (**b**) compared with the original data (**a**)

From Fig. 6.11, we can see that interpolation result contains the basic information of original data and can better reflect the details.

To understand the accuracy of the interpolation method more deeply, we compare the original data with interpolation data in slice. The result is displayed in Fig. 6.12.

From Fig. 6.12, we can see that in X, Z direction, the original data and interpolated data are almost the same; in Y direction, however, the original and the interpolated data have the same shape, but the data are not matching well. This needs to be improved.

Finally, we make an evaluation on the effectiveness of the interpolation method. In this evaluation progress, we use percentage of average error estimates (PAEE), i.e.

$$\text{PAEE} = \frac{100\ \%}{Z_n} \sum_{k=1}^{n} \left[Z_i^*(x_k) - Z_i(x_k) \right]^2, \tag{6.42}$$

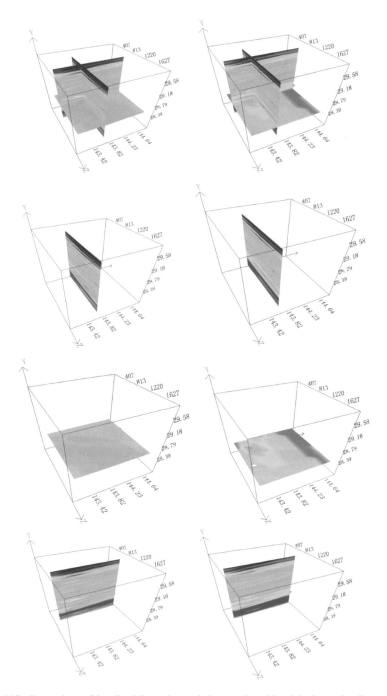

Fig. 6.12 Comparison of localized linear interpolation results with original ones in slices

where x_k is position, Z_i^* is the estimated value of random variable Z_i at x_i, $Z_i(x_k)$ is the sample value at x_k, Z_n is average value of all samples, n is the number of samples. If PAEE $\rightarrow 0$, the estimation is unbiased. After calculation, the PAEE of this method is 1.07 %. This is a quite small number, meaning that this method is practically effective.

Chapter 7
Imaging of Theoretical Model and Field Examples

To better understand pseudo-seismic interpretation and migration imaging technology in TEM method, numerical results on migration imaging of TEM theoretical models and examples from field data are presented.

7.1 Model Calculations

7.1.1 Layered Model

Figure 7.1 shows the working arrangement at the earth surface. There are 11 profiles with 11 measuring points in each profile. The side length of transmitter loop is 200 m and the width of sampling window ranges from 32.5 μs to 8.7 ms. All of the 121 measuring points are forward calculated. Using the wave field conversion, we converted the electromagnetic responses obtained into pseudo-seismic field values at the survey points to calculate the wave field extension value of geoelectric section.

The parameters of a three-layer A-type geoelectric model are as follows: $\rho_1 = 50\,\Omega\,\text{m}$, $\rho_2 = 500\,\Omega\,\text{m}$, $\rho_3 = 5000\,\Omega\,\text{m}$, $h_1 = 100\,\text{m}$, $h_2 = 200\,\text{m}$. We calculated forward model electromagnetic responses for each measuring point and then carried out wave field conversion. For clarity, only the converted crest is displayed, shown in Fig. 7.2. From the figure, it is shown that with increasing virtual time τ, the virtual wave field demonstrates two positive peaks, which are associated with A-type geoelectric model. Using electromagnetic responses from each measuring point, virtual velocity values are calculated by $v = \frac{1}{\sqrt{\mu_0 \sigma}}$ and results are shown in Fig. 7.3. The velocity variations are reflected in three distinct layers as the virtual values of wave field depends on subsurface conductivity which subsequently corresponds to the distribution of the model's resistivity.

© Science Press and Springer Nature Singapore Pte Ltd. 2017
X. Li et al., *Migration Imaging of the Transient Electromagnetic Method*,
DOI 10.1007/978-981-10-2708-6_7

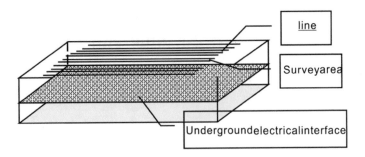

Fig. 7.1 Schematic diagram of calculation model

Finally, the three-dimensional boundary element technique is applied to the Kirchhoff integral calculation, together with the use of the velocity analysis, a three-dimensional continuation imaging is shown in Fig. 7.4 having two underground positions of A-type geoelectric model. The corresponding depths of the two positions are also very close to 100 and 300 m which are consistent with the model parameters.

7.1.2 Three-Dimensional Model

In order to verify the effectiveness of the method presented in this book, we designed three-dimensional models with a resistive abnormal body and a conductive abnormal body embedded in a half-space. The transmitter is a central loop

Fig. 7.2 The wave field for an A-type geoelectric model

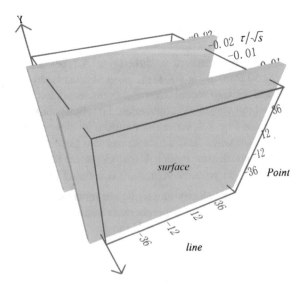

Fig. 7.3 The velocity analysis for the A-type geoelectric model

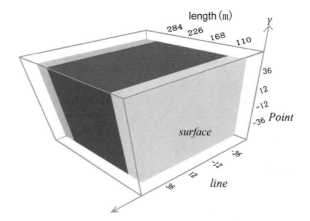

Fig. 7.4 The wave field continuation result for the A-type geoelectric model

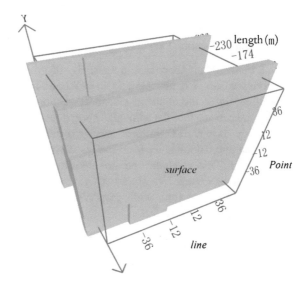

configuration. Taking the EM responses to the wave field transformation, we obtain the synthetic wave field through synthetic aperture calculation. Finally, the three-dimensional boundary element method is used for Kirchhoff migration imaging.

As shown in Fig. 7.5a, the abnormal body is a block of 30 m × 30 m × 50 m. The side length of the transmitter is 100 m. The measuring points at the earth surface are shown in Fig. 7.5b. The main profile has 11 measuring points with interval of 10 m.

(a)

(b)

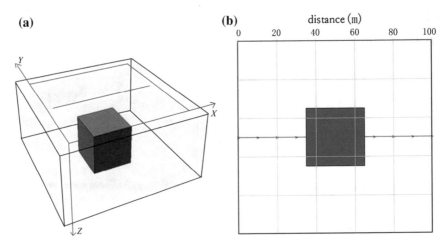

Fig. 7.5 a Three-dimensional model and its plane view (**b**)

(1) A resistive anomalous body

Take a uniform half-space with a resistivity of $\rho_1 = 10\,\Omega$ m. The cubic body has a resistivity of 300 Ω m and a top depth of 70 m. First, the forward modeling method is used to obtain the apparent resistivity section as shown in Fig. 7.6. From the figure, one can see that the anomaly center does not correspond to the center of the cubic body. It is shifted upward, making it difficult to determine the location and size of the anomalous body from the resistivity section.

Using the TEM synthetic aperture algorithm, the data of 11 points of the model are calculated. Figure 7.7a shows the results of TEM wave field transformation before synthetic aperture, while Fig. 7.7b shows the results of TEM wave field

Fig. 7.6 Apparent resistivity section diagram

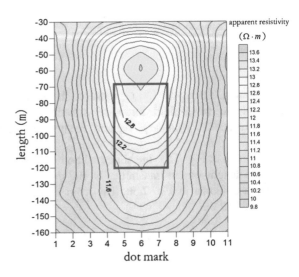

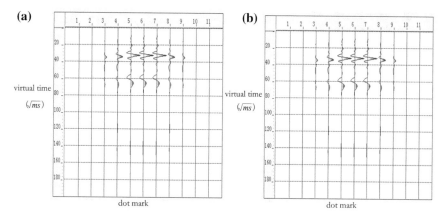

Fig. 7.7 Model results. **a** TEM wave field transformation before synthetic aperture; **b** TEM wave field transformation after synthetic aperture

transformation after synthetic aperture. Comparing the two figures, we can see that the synthesized waveforms obtained certain enhancement in the central part with good correlations. The waveforms weaken in the edge portion, where the correlations are poor. The effectiveness of the synthetic aperture is verified.

Finally, we can obtain the imaging results on the main cross-section after the speed analysis and Kirchhoff migration of wave field, as shown in Fig. 7.8. From the figure, one can see that the two peaks—the upper and the lower peak—are at 70 and 120 m, respectively. This has a good agreement with model. The vertical resolution improves as well. Waveform amplitude of the second interface reduces a lot compared to first interface. This is caused by the weak reflection at the surface. Comparison with the above model of a resistivity cube, the present model has a higher resolution.

(2)   A conductive anomalous body

We assume that a conductive cubic body with a resistivity of $\rho_1 = 5\,\Omega\,m$ located in a uniform half-space of $\rho_0 = 25\,\Omega\,m$. The dimension of cube remains unchanged, but its depth increases to 120 m. The apparent resistivity section on the main cross-section is shown in Fig. 7.9. From the figure, one can see that the center of the low-resistivity anomalies does not correspond to the position of the model and there is a little downward-shift. The size and location of the abnormal body also appear to be deviated in the resistivity profile. Through wave field transformation, we can get the comparison of main cross-sections before and after synthesis, as shown in Fig. 7.10. Similarly, the waveforms show better resolution in the central part with good correlations, while poor resolution away from center, where correlation is poor which validates the effectiveness of the synthetic aperture.

Figure 7.11 shows the result of migration imaging. It is inferred that structure consists of two interfaces. Because the depth of the abnormal body increases and

Fig. 7.8 Imaging section of migration along main profile

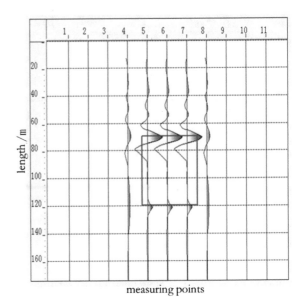

Fig. 7.9 Apparent resistivity section

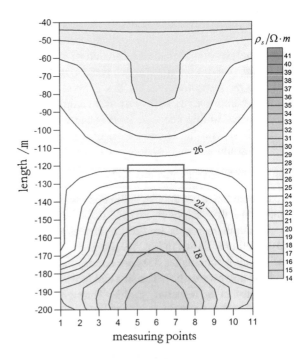

the waves of the low-resistivity model broaden, the position moves slightly downward.

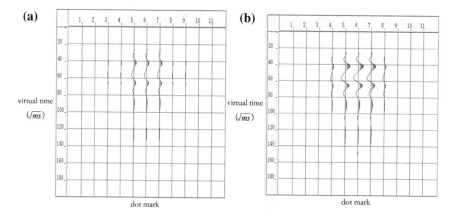

Fig. 7.10 The model results for **a** TEM wave field transformation before synthetic aperture; **b** TEM wave field transformation after synthetic aperture

Fig. 7.11 Imaging section of migration along main profile

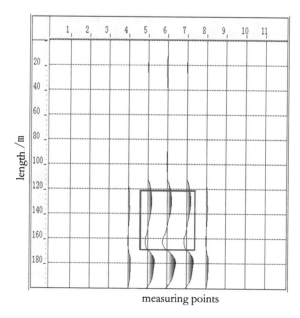

7.2 Examples with Field Data

7.2.1 Advanced Detection of Tunnel

We first apply the technique presented in this book to tunnel advanced detection in Sichuan province, China with in a complex geological environment. The auxiliary hole project is located on a slope area from the Qinghai-Tibet plateau to the Sichuan

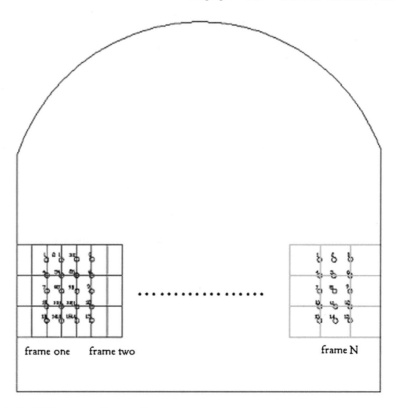

Fig. 7.12 TEM layout and survey lines

basin. The terrain is characterized by rough topography and elevations are all above 3000 m. The highest peak is 4125 m. Part of the hole has a length of about 12,875 m and a depth of more than 1500 m, which takes up 73.1 % of the complete hole. The auxiliary hole of Jinping hydropower project is located at big Jinping river bend, main stream of the Yalong river, the junction area of Muli, Yanyuan, Mianning county in Liangshan Yi Autonomous Prefecture, Sichuan province. To ascertain water in the auxiliary hole, in different tunnel faces of the auxiliary hole, four survey profiles were arranged [A tunnel face (mileage AK11 + 843), A tunnel face (mileage AK11 + 447), A tunnel face (mileage AK11 + 598), the cross-channel (# 7-2) cross-section]. Through detection experiments to the auxiliary hole of Jinping hydropower project, we try to verify the effectiveness of the method.

The survey lines for the advanced tunnel forecast using TEM method are shown in Fig. 7.12. Five lines are measured in each wireframe with each line measuring 3–5 points, and then the frame is moved. We repeated these steps until entire tunnel face survey is completed. Multiple test lines can not only ensure adequate information, but also make sure that the source data can define the target from the aspect of continuous velocity analysis and continuation imaging.

Fig. 7.13 Velocity
distribution of virtual wave
field

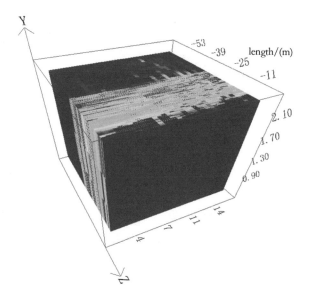

We first calculate a set of initial virtual velocities from the measured electromagnetic responses, and then conducted three-dimensional interpolation twice for
velocity analysis. The results of velocity analysis are shown in Fig. 7.13. From the
figure, we can see that the distribution of virtual velocity approximately demonstrates three layers. The values change at 7 and 30 m, corresponding to the real
resistivity distribution. This demonstrates that the virtual velocity can well reveal
the real resistivity distribution in the underground.

Next, we make the wave field transformation of the collected data and displayed
the results in Fig. 7.14. From the figure, we can see that the characteristics of the
wave field with a number of peaks and troughs are demonstrated.

The results of wave field transformation are displayed in three-dimensional form,
where only the part connected with the water body retains. As can be seen from
Fig. 7.15, the layer interfaces and their morphology are clearly observed.

Similarly, the continuation results are displayed in three-dimensional form. As
can be seen from Fig. 7.16, the positions around 7 m and from 20 to 35 m reflect
spatial distribution of water.

Finally, by comparing with the known situations of tunnel face in the front, we
can assure that the results after a series of wave field transformation, velocity
analysis, and continuation imaging coincide well with real geological conditions.

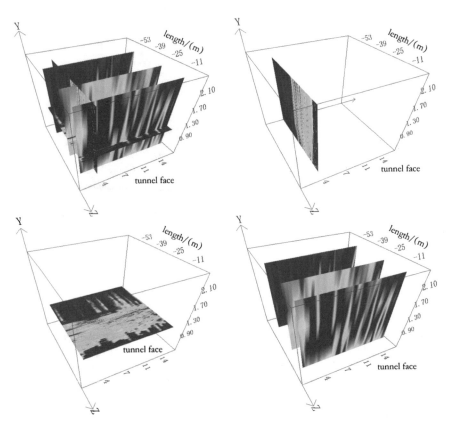

Fig. 7.14 Sliced display of the virtual wave field from the survey data

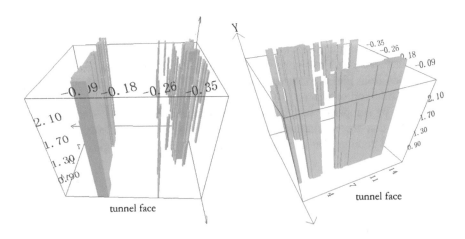

Fig. 7.15 Wave field transformation results of water body viewed from different directions

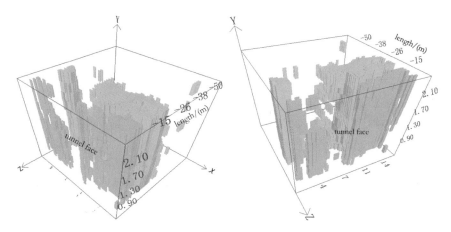

Fig. 7.16 Wave field continuation results of water body

7.2.2 Detection of Goaf of Coal Mine

7.2.2.1 Example 1

In the process of coal mining, water disaster, such as goaf and collapsing of column has challenged the coal production. The purpose of investigations in a coal mining area is to determine the depth, height, and the distribution of the goaf filled with water and to avoid water inrush disaster in the mining process.

The detection area is mainly covered by loess of Quaternary with thickness generally not more than 20 m. The underlying strata include mainly the Tertiary sandy conglomerate layer, the sand-shale with coal seams of Permian, the sand-shale with coal seams of Carboniferous age and Ordovician limestone. The main coal seams in this area have two layers, located at 150 and 260 m, respectively. It is difficult to cause significant resistivity anomaly because the resistivity variations are small. When the underground coal is mined out, the local stress concentrates locally. The gob roof under the influence of overburden pressure forms deformation, fracture, displacement, caving, and caving zone. With the filling of groundwater and surface water along the crack to the mined-out area, the resistivity will change obviously, forming a low-resistive body. This provides the favorable condition for the detection of mined-out area.

Based on the buried depth and thickness of the mined-out area near the work area and combined with the characteristics of TEM, we chose a fixed loop configuration. The side length of the loop transmitter is 200 m, the line spacing is 20 m, the interval between measuring points is 20 m.

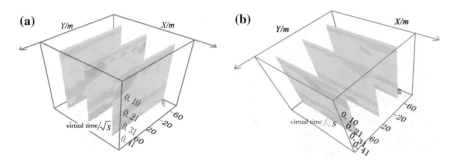

Fig. 7.17 Three-dimensional view of wave field. **a** Section in X-direction. **b** Section in Y-direction

(1) Wave field transformation

Figure 7.17 is three-dimensional imaging after wave field transformation, the equalization processing, deconvolution, and smoothing. The dark color is the area of high virtual wave energy, which is also the location of reflection interfaces.

(2) Wave field continuation imaging

To better display the spatial distribution of the goaf, we present wave field continuation imaging results in two ways. One is to get rid of the reflection part of low energy (low speed or low resistance), as shown in Fig. 7.18a. It clearly shows the positions of the mined-out area in the rocks. The other is to get rid of high-energy reflection (high speed or high resistance), as shown in Fig. 7.18b that displays the spatial distribution of the mined-out area. Through analysis of the imaging results, we find two-layer mined-out area. The upper goaf is located 140 m deep, which is filled with water and the caving zone is small; the lower mined-out

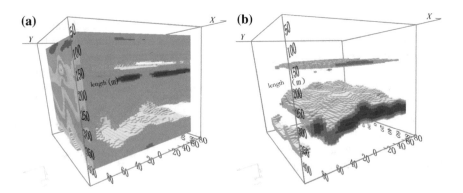

Fig. 7.18 The imaging result of three-dimensional wave field continuation. **a** Without low energy reflections. **b** Without high-energy reflections

area is located at 260 m deep, which is filled with water and the caving zone is large. Because of the water-filled effects, the low-resistivity anomaly is enlarged and distribution of mined-out area is also amplified. The interpretation results are in good agreement with the actual geological conditions. Therefore, field data demonstrates the effectiveness of the algorithm.

7.2.2.2 Example 2

For a lot of coal mines in Shanxi Province, China, the geological disasters such as seeper, goaf, collapsing of the columns pose serious threat to the coal mining. Therefore, it is necessary to properly investigate the coal mine goafs by TEM method.

The instruments used in the work are GDP—32II electrical method workstation. Considering the geology of the working area and the target, we chose TEM and the fixed loop device. The side length of the loop transmitter is 300 m, the spacing between survey lines is 20 m, the interval between measuring points is 20 m. The receiver uses special TEM probes (reception area is 2000 m^2). The transmitting frequency is 25 Hz and the time ranges 0.087–8 ms.

Figure 7.19a shows the apparent resistivity distribution. Two low-resistivity anomaly zones appear in northwest and southeastern part caused by water-filled goaf. Apparent resistivity distribution of the map indicated two main low-resistivity anomaly zones (A and B) characterized by contour of 65 Ω m. The red square box in Fig. 7.19a is the data collection area for three-dimensional imaging. Figure 7.19b shows the cross-section of apparent resistivity along main profile. The left side of the red line is the data used for three-dimensional imaging. From the section, we can see that the resistivity is higher in the shallow area, reaching more than 100 Ω m. From 120 m, the resistivity decrease rapidly and reaches minimum of about 55 Ω m at 150 m. Then, the resistivity increases with depth. At 500 m, it is back to 100 Ω m. From the figure, two electrical interfaces are clear. The first layer is located at the depth of 130–160 m, while the second layer interface is difficult to determine ranging from 250 to 350 m.

We combine the data of point 300–500 from line 1300 to 1500 in the working area to form a three-dimensional data volume. Then, we can get a three-dimensional imaging of the wave field in the entire region after the wave field transformation, appropriate equalization processing, and synthetic aperture calculation. Figure 7.20 shows that correlation superposition improved the signal-to-noise ratio.

We performed the migration imaging of the synthetic data to get the three-dimensional imaging. The vertical axis represents depth, as shown in Fig. 7.21. The blue color is for the negative wave field, while the red color is for the positive wave field; the yellow and green are the transition colors, showing the small wave field. From the figure and compared with a sectional view of resistivity we can see that the layering is very clear. Through the interpretation of the results, we believe that there is a two-layer goaf. The upper gob roof is at 140 m deep; the

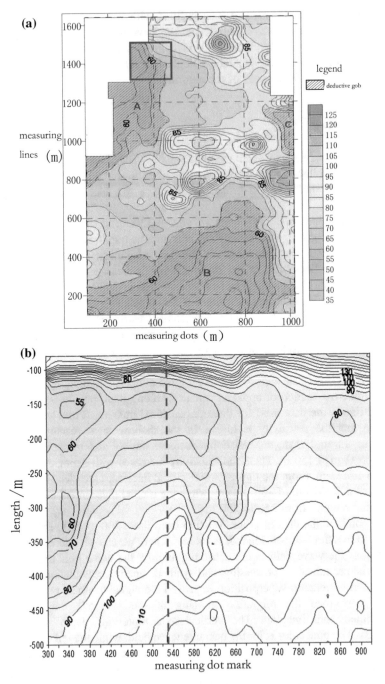

Fig. 7.19 The apparent resistivities of the working area. **a** Plane view of the apparent resistivity. **b** Typical apparent resistivity section

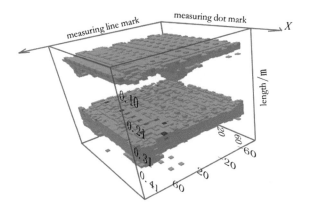

Fig. 7.20 Three-dimensional imaging of typical section

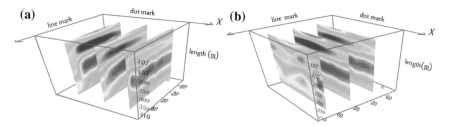

Fig. 7.21 Three-dimensional imaging of wave field. **a** Section in X-direction. **b** Section in Y-direction

lower gob roof is at 280 m deep. Because of the water-filling effects, the low-resistivity anomaly gets enlarged and the distribution of mined-out area also gets enlarged. The interpretation results are in good agreement with the actual geological situations. The effectiveness of the algorithm is again proved.

Printed in the United States
By Bookmasters